COCKPIT DISPLAYS: TEST AND EVALUATION

To the memory of Jim Newman, 1967-1986

Oh, I have slipped the surly bonds of earth
And danced the skies on laughter-silvered wings;
Sunward I've climbed, and joined the tumbling mirth
Of sun-split clouds — and done a hundred things
You have not dreamed of — wheeled and soared and swung
High in the sunlit silence. Hov'ring there
I've chased the shouting wind along, and flung
My eager craft through footless halls of air.
Up, up in the long, delirious, burning blue
I've topped the windswept heights with easy grace
Where never lark, or even eagle, flew,
And, while with silent, lifting mind I've trod
The high untrespassed sanctity of space,
Put out my hand, and touched the face of God.

John Gillespie Magee, Jr.
High Flight

Cockpit Displays: Test and Evaluation

RICHARD L. NEWMAN
Embry Riddle Aeronautical University
Prescott, Arizona, USA
KEVIN W. GREELEY
Lockheed Martin
Marietta, Georgia, USA

Routledge
Taylor & Francis Group

LONDON AND NEW YORK

First published 2001 by Ashgate Publishing

2 Park Square, Milton Park, Abingdon, Oxon OX14 4RN
711 Third Avenue, New York, NY 10017, USA

Routledge is an imprint of the Taylor & Francis Group, an informa business

First issued in paperback 2016

British Library Cataloguing in Publication Data
Newman, Richard L.
    Cockpit displays: test and evaluation
    1.Aeronautical instruments - Display systems
    I.Title II.Greeley, Kevin W.
    629.1'35

Library of Congress Control Number: 00-109132

ISBN 978-0-7546-1549-1 (hbk)
ISBN 978-1-138-25809-9 (pbk)

Transferred to Digital Printing in 2014

Contents

List of Figures

List of Tables

Acknowledgments

It is difficult to identify the many colleagues who have influenced the ideas presented in this book. We have indeed been fortunate in the sheer number of individuals who have helped us along the way. To those whom we omit, please accept our apologies and know that we appreciate your efforts.

There are, however, a few who deserve mention by name. The many interesting and often heated discussions with Loran Haworth of Ames Research Center, Ron Kruk of CAE Electronics, Terry Turpin of Turpin Technologies, and Ryan Wilkins of Boeing-Philadelphia helped focus our thoughts on display evaluation in rotorcraft.

Richard Schwartz of Lockheed-Martin provided valuable insight into optical test and evaluation and on human vision limitations. Eugene Adam of McDonnell-Douglas (now Boeing) was a valued mentor.

Ryan Wilkins of Boeing-Philadelphia and Malcolm Burgess of Research Triangle Institute reviewed some material and provided comments. We appreciate this help.

Abbreviations

1553	MIL-STD-1553
429	ARINC 429
A-320	Civil Transport, Airbus
A-330	Civil Transport, Airbus
A-340	Civil Transport, Airbus
A-7	Military Attack Airplane, Vought *Corsair II*
A/D	Analog to Digital
AC	Advisory Circular
accel/decel	Acceleration/Deceleration
ACIDTEST	Aircraft Cockpit Information Display Tenets Expert System Tool
ACM	Air Combat Maneuvers
ADF	Automatic Direction Finder
AFB	Air Force Base
AFFSA	Air Force Flight Standards Agency
AFM	Approved Flight Manual
AH-1	Military Helicopter, Bell *Cobra*
AH-64	Military Helicopter, Boeing (née McDonnell-Douglas, née Hughes) *Apache*
AHRS	Attitude/Heading Reference System
ARINC	Aeronautical Radio, Inc.
ARP	Aerospace Recommended Practice
ASW	Anti-Submarine Warfare
ATC	Air Traffic Control
ATK	Attack
ATM	Aircrew Training Manual
AWACS	Airborne Warning and Control System
B-727	Civil Transport Aircraft, Boeing
B-747	Civil Transport Aircraft, Boeing
B-757	Civil Transport Aircraft, Boeing

BMR	Bomber
CAA	Civil Aviation Authority
CAR	Cargo or Transport
CCD	Configuration Control Document
CDI	Course Deviation Indicator
CDM	Climb-Dive Marker
CFIT	Controlled Flight into Terrain
CLSA	China Lake Situation Awareness
CMS	Cockpit Management Systems
COTS	Commercial Off-the-Shelf
CRM	Crew (or Cockpit) Resource Management
CRT	Cathode Ray Tube
CSA	Crew Situation Awareness
CSDD	Crew Station Design Document
CTR	Civil Tilt Rotor
DEP	Design Eye Position
DFR	Display Flyability Rating
DGPS	Differential Global Positioning System
DH	Decision Height
DME	Distance Measuring Equipment
DoD	Department of Defense
DRR	Display Readability Rating
DT&E	Developmental Test and Evaluation
EEG	Electroencephalogram
EFA	European Fighter Aircraft
EGPWS	Enhanced Ground Proximity Warning System
EID	Ecological Interface Design
EMC	Electromagnetic Compatibility
EMD	Engineering/Manufacturing Development
EMI	Electromagnetic Interference
F-15	Military Fighter, Boeing (née McDonnell-Douglas) *Eagle*
F-16	Military Fighter, Lockheed (née General Dynamics) *Fighting Falcon*

F-18	Military Fighter, Boeing (née McDonnell-Douglas) *Hornet*
FAA	Federal Aviation Administration
FADEC	Full Authority Digital Engine Control
FAR	Federal Aviation Regulation
FBW	Fly by Wire
FCS	Flight Control System
FFT	Fast Fourier Transform
FL	Flight Level
FMEA	Failure Modes and Effects Analysis
FMS	Flight Management System
FPM	Flight Path Marker
FQ	Flying Qualities
FTR	Fighter
FW	Fixed-Wing
FWTCI	Frequency Weighted Task Complexity Index
GCAS	Ground Collision Avoidance System
GPS	Global Positioning System
GPWS	Ground Proximity Warning System
GS	Glideslope
H	High Power NDB, useful distance 50 nm
HDD	Head-Down Display
HERF	High Energy Radiative Field
HH	High Power NDB, useful distance 75 nm
HMD	Helmet-Mounted Display (or Head-Mounted Display)
HQ	Handling Qualities
HQR	Handling Qualities Rating
HSCT	High Speed Civil Transport
HTS	Head Tracker System
HUD	Head-Up Display
IAW	In Accordance With
IFR	Instrument Flight Rules
IHADSS	Integrated Helmet and Display Sighting System
ILS	Instrument Landing System
IMC	Instrument Meteorological Conditions

IP	(1) Initial Point: (2) Instructor Pilot
JAA	Joint Aviation Authority
JAR	Joint Airworthiness Requirements
KIAS	Knots, Indicated Airspeed
KISS	Keep It Simple, Stupid
LASER	Light Amplification by Stimulated Emission of Radiation
LIDAR	Light (or Laser) Detection and Ranging
LOS	Line of Sight
LOSA	Loss of Situation Awareness
MAIR	Mission Analysis and Information Requirements
MD-80	Civil Transport, Boeing (née McDonnell-Douglas)
MDA	Minimum Descent Altitude
MEL	Minimum Equipment List
MH	Medium Power NDB, useful distance 25 nm
MIDAS	Man-Machine Integration Design and Analysis System
MIL-STD	Military Standard
MMI	Man-Machine Interface
MRT	Modified Rhyme Test
MTBF	Mean Time Between Failures
MTE	Mission Task Element
NASA	National Aeronautics and Space Administration
NDB	Non-Directional Beacon
NIH	Not Invented Here
NOE	Nap-of-the-Earth
NR	Not Recommended
NT-33	Research Airplane, Lockheed *T-Bird*
NTSB	National Transportation Safety Board
NVG	Night Vision Goggles
OFE	Operational Flight Envelope
OP	Operational Pilot
OT&E	Operational Test and Evaluation
PCP	Proximity Compatibility Principle
PFD	Primary Flight Display
PNF	Pilot Not Flying

POC	Proof-of-Concept
RA	Resolution Advisory
RAH-66	Military Helicopter, Sikorsky-Boeing, *Comanche*
REC	Reconnaissance
RTCA	Radio Technical Committee for Aeronautics
RW	Rotary-Wing
S-by	Standby (Display)
SA	Situation Awareness
SA-SWORD	Situation Awareness Subjective Workload Dominance
SAE	Society of Automotive Engineers
SAGAT	Situation Awareness Global Assessment Technique
SAP	Situation Awareness Probe
SAR	Search and Rescue
SART	Situation Awareness Rating Technique
SAS	Stability Augmentation System
SASRF	Situation Awareness Supervisory Rating Form
SCT	Scout
SDO	Spatial Disorientation
SFE	Service Flight Envelope
SHCT	Short Haul Civil Transport
SKE	Stationkeeping Equipment
SME	Subject Matter Expert
SOP	Special Operations
SWAT	Subjective Workload Assessment Technique
SWORD	Subjective Workload Dominance
T	Standard Arrangement of Flight Instruments (also known as Basic T)
T&E	Test and Evaluation
TA	Traffic Advisory
TACAN	Tactical Air Navigation
TAWS	Terrain Awareness Warning System
TCAS	Traffic Alert and Collision Avoidance System
TERPS	Terminal Instrument Procedures
TF/TA	Terrain Following/Terrain Avoidance
TKR	Tanker

TLAR	That Looks About Right
TLX	Task Load Index
TP	Test Pilot
TSO	Technical Standard Order
UA	Unusual Attitude
USAF	United States Air Force
UTL	Utility
V-22	Military VTOL Aircraft, Bell-Boeing *Osprey*
VAPS®	Virtual Avionics Prototyping System®
V_C	Corner speed, i. e. stall speed at design load factor
VCR	Video Cassette Recorder
V_H	Level Flight Speed at Maximum Continuous Power
VHF	Very High Frequency
VOR	VHF Omnirange
VORTAC	Collocated VOR and TACAN Station
VTOL	Vertical Takeoff and Landing
WAD	Workload Assessment Device
WBS	Work Breakdown Structure

1 Introduction

The Need for Integrating Design and Test and Evaluation

There have been a number of papers and articles written about operational difficulties with modern display and other cockpit systems.* As we see it, the problem has been a series of discontinuities between the users and the designers, between the designers and the testers, and between the users and the testers.

As a result of the first discontinuity, between users and designers, inadequate design requirements are established. This is particularly unfortunate as systems can be (and are being) designed with greater and greater capabilities in terms of automatic flight and guidance and flight control. Without adequate requirements, it is hardly surprising that there are problems encountered in operational use.

The second discontinuity, between designers and testers, reduces the opportunities for feedback to the designer. In fact, with today's economic setting, many systems are practically committed to production by the time they reach flight test. Only if there is very serious problems, will these systems be corrected.

The third discontinuity, between users and testers, results in inadequate test criteria. As a result of this discontinuity, we are left with highly subjective criteria which vary from tester to tester. Or we have inappropriate criteria.

Essential Features of a Test and Evaluation Philosophy

The essential features of our test and evaluation (T&E) philosophy are easy to state.

➔ Feedback to the design team

---

* These operational difficulties will be covered later in Chapter 2 (page 4) and Chapter 7 (page 53)

1

- ✈ Valid and objective test completion criteria
- ✈ The ultimate tester is the user
- ✈ We are testing displays, not pilots

Feedback to the Design

There should be prompt and early feedback to the design team. The development process should allow for early proof-of-concept designs with early test feedback. This is the only way we will obtain improved systems which meet the needs of the user.

Objective Test Completion Criteria

There should be valid and objective test completion criteria. The use of subjective criteria should be avoided. We feel that objective test completion criteria makes the ultimate procurement specification. The objective criteria must be based on performance measurements. The final tests (and to some extent the important tests) are mission scenarios and situation awareness tests. All tests to that point are for risk reduction or proof-of-concept.

This is important enough to repeat... *The test criteria should be based on objective performance measures.*

Ultimate Testers

The ultimate tester is the user. A frequent question is "Test pilots or operational pilots?" We strongly believe that the objective performance measures mentioned above must be achieved by operational pilots — the users. For this reason, we feel that test pilots should not act as evaluators during mission and situation awareness testing.

Final mission testing should use customer pilots. Test pilots should certainly participate in this testing, but as test directors, not as evaluators. If the test pilot community has concerns with aspects of a cockpit system, they need to design test scenarios to address these concerns.

Testing Displays, not Pilots

We are testing displays, not pilots. Finally, the purpose behind this book it articulating better ways of evaluating displays and related cockpit systems. While some of the methods and techniques discussed can be used for psychological testing of human subjects, the primary purpose of the discussion is how best to ensure that our displays and cockpit systems aid in the performance by the using pilot — at the sharp end of the airplane.

Statement of the Problem

Put to its simplest form, the problem is lack of requirements in many design efforts, lack of test requirements, and minimal feedback to the designer.

2   Lessons Learned

Cockpit Automation

The most common question in modern cockpits is reported to be "What's it doing now?" There are anecdotal reports of pilots turning the automated system off when air traffic control (ATC) changes the landing runway because it's easier to do without the "help" from the automation than to reprogram it.*

Cockpit automation has come under fire recently because of a series of operational incidents.(*1-2*)† Mode confusion was reported to be a serious problem by the NTSB. Increasing mode complexity is prompting one operator to modify its current flight management systems (FMSs) to reduce the number of operating modes.

Recently, a study by the Australian Bureau of Air Safety Investigation published a report (*3*) on a survey of Pacific rim pilots flying FMS-equipped airplanes. Among the troubling results are the numbers of design induced errors (mode selection and transposition of heading-select and course-select functions) and the large number of "work-arounds" necessary to manipulate the FMS. These are most often reported during descents. It was not clear if these were the result of incompatible ATC procedures or of poor software design.

The problem seems to be one of pilots frequently suffering from a loss of situation awareness (LOSA) because of the increasing complexity of FMS modes. A recent review article was titled "How in the World Did We Ever Get into That Mode?"(*4*) This article, by Sarter and Woods, suggests that current cockpit automation makes it more important and more difficult for pilots to remain aware of the status of the system's different modes of operation. They cite research where pilots made critical errors during non-standard situations, such as leaving autothrottles engaged during aborted takeoffs.

---

* T. G. Foxworth (United Airlines), personal communications, 1992-1994.

† Italicized numbers in parentheses denote references listed beginning on page 258.

The likelihood of mode errors increases when the operating rules change from mode to mode.(5) This would be true of the mode changes are not obvious or well-annunciated. Sarter and Woods cite the FCS in a modern fly-by-wire (FBW) transport which has a two vertical modes which differ in their speed control methods. Of these, the non-standard OPEN DESCENT mode can be entered inadvertently and was thought to have been a factor in an approach accident.(6)

Billings (7) observes that present automation reduces the workload during normal operations but increases it during abnormal operations. He argues that the systems should be designed to reduce workload and error during abnormal or emergency operations. The normal operational case is well within pilot capabilities.

Billings (8) also suggests that cockpit automation should adequately convey status information to the flight crew and the systems should be accountable, comprehensible, and informative.

Some electronic attitude indicators remove unneeded information, such as mode annunciation, during extreme attitudes — a process designed to enhance the ability of the pilot to use the display for recovery without distraction. This removes mode awareness from the pilot and was thought to have contributed to the accident to the Airbus A-330 at Toulouse.*

Cockpit Displays

Symbol Choices

With any electronic aircraft display, head-up, head-down, or helmet-mounted, there are two divergent forces. On the one hand, there is a great clamor for standardization of symbology. At the same time, there is an extraordinary desire to make every aircraft application different. Any student of head-up display (HUD) history will testify to this.

Electronic displays can be developed in almost any format. In spite of this, they have often mimicked existing conventional panel instruments. Similarly, HUD symbology often mimics head-down displays. This has resulted in confusion over control techniques, in excessively clut-

---

* H. B. Green (FAA, Seattle), personal communication, September 1994.

tered displays, and in displays which do not make the best use of the capabilities of the systems.

Similarly, some proposed HMD symbology formats appear to be copied inappropriately from HUD symbologies.

Lack of Criteria

What has been lacking is any organized set of development, test, and evaluation criteria for displays. As a result, HUD development usually progresses through a series of personal preference choices by either the manufacturer's project pilot or the customer's pilot.

As decisions are made, the rationales for the choices aren't documented, forcing new systems to go through the same process over and over.

Symbol Control Laws

Control laws and algorithms which drive the various symbols have not been well described. The absence of specifications and of documentation has created problems with HUDs where the symbols were excessively noisy (lateral motion of the F-16A FPM) or led to pilot uncertainty about the origin of the data (aircraft reference symbol in the MD-80).

Historically, there have been no requirements to deliver the display code as part of the data package. This makes it quite difficult to determine exactly what is displayed and how the symbols are driven. Manufacturers treat the source code as proprietary data.

Integration

Many display systems, particularly HUDs, are installed as add-ons. If inadequate attention is paid to integrating the HUD with existing systems, excessive pilot workload can result. This may not be apparent in most situations, but can become overwhelming with a small addition to external workload. In a recent flight test,(9) poor system integration did not become apparent until operational trials. The difference between various ATC workloads resulted in a display being rated as *satisfactory* during low workload situations and *unacceptable* when, for example, the pilot was asked to "maintain 180 knots to the marker" and vectored through the localizer before final intercept.

Software Validation

A major constraint is the need to validate the software which performs the algorithms driving the symbols. This can require a considerable amount of time. Usually the validation is well underway before the display evaluation is begun. As a result, there is an extreme reluctance to modify any symbol or control law since it will require revalidation and a large increase in cost. It is often said that there is no such thing as "changing one line of code."

The display symbology thus becomes frozen before test and evaluation. It is expensive to change even a minor item, such as the shape of a symbol, not because of the effort to make the change, but because of the lengthy validation and verification of the software. The use of literals, rather than variables, makes the software changes even more intractable.

Clutter

Frequently, in the absence of design criteria or a valid methodology, more and more information is added to the display, "because we can." This leads to an excessively cluttered presentation. The problem can reach such a level that declutter logic is required to provide a usable display during critical flight phases, such as recovery from unusual attitudes. This is usually thought of as a HUD/HMD problem, but it has created problems with head-down displays as well.

Minimizing clutter is a paramount issue for see-through displays to allow the pilot to see real-world objects.

Fly-By-Wire

Many modern aircraft are being designed with fly-by-wire (FBW) flight control systems (FCSs). FBW means there is no direct connection between the pilot's controls and the aircraft control surfaces. This allows the air vehicle's response to control inputs to be tailored to provide good flying qualities independently of airspeed or configuration. FBW also makes it possible to separate the stability and control characteristics from the geometry of the vehicle. This allows stealthy designs, extremely maneuverable fighters, and the reduction of the size (and drag) of the empennage.

There are several human factors issues related to FBW systems. These are control authority, control feedback, and control modes.

Control Authority

One of the advantages of FBW design is the ability to protect the vehicle from the pilot. A full authority digital engine control (FADEC) can be designed to control engine thrust directly and not some indirect parameter, such as air flow. The FADEC can also stop commanding more thrust when the engine operating limits are reached. This is normally considered to be a *good idea*.

Digital FCSs can also be designed to prevent the pilot from exceeding the stall angle-of-attack, limit airspeeds, or the design load factor. These limits are also considered by the designer to be *good ideas*. But are they?

Flight crews are not convinced that these hard limits are desirable. They argue that it may be preferable to overstress the airplane when the alternative is hitting the ground. It may be better to overtemp the engine during a windshear than the alternative of hitting the ground.

The designers counter with the argument that having hard limits encoded in the controls allows the pilots to reach the limits faster and will actually improve the response.* At this writing, neither side has provided convincing arguments.

Control Feedback

Part of the pilot's cues are the tactile feedback from the controls themselves. This has been recognized for some time. The airworthiness requirements specify that elevator, aileron, and rudder control forces must reflect the response of the airplane.(*10-11*) However, additional cues are useful to the pilot: control inputs from the other pilot, from the autopilot, from the trim system, and from the autothrottles. Some modern airplanes have been designed, for example, with autothrottles which do not move to indicate changes in thrust commanded by the automatic system. Is this feedback necessary? Perhaps not, but it does deprive the pilot of an addi-

---

* Gordon Corps (Airbus Industrie), comments made at Society of Automotive Engineers S-7 committee meeting, Stockholm, May 1985.

tional cue in an airplane where he is already having difficulty keeping track of the FMS modes.

A similar comment can be made for the need for feedback to one pilot of the other's control inputs. Historically, each pilot could monitor the other's intentions by noting the control input. One of the rules for human-centered design is that each element of the system must have knowledge of the other's intent.(8)

Control Modes

Difficulty in remaining aware of the various FCS modes follows the same discussion as in the earlier section on automation modes. The impact may be more critical because of the more critical nature of flight controls but also because FCS mode changes are more likely to be made without direct pilot intervention. FCS mode changes can be caused by configuration changes, by changes in the environment, or by equipment failures. Because of this, the pilot may not be aware of the changes unless they are clearly annunciated.

The common thread of these FBW issues is the lack of feedback from the operating crews to the design team.

Other Systems

A good example of needless complication, exacerbated by gold-plated specifications is the current design of global positioning system (GPS) receivers for civil aviation. Many relatively inexpensive GPS receivers are equipped with a world-wide database of navigational radio beacons, air traffic control (ATC) fixes, airports, etc. The ability to prepare multiple flight plans, store hundreds of user-generated fixes, and record many flight functions make the design of the interface daunting. Yet, in our experience, most of these features are simply too complicated to use, like the design of most VCRs.* Most users simply want to go from point A to point B with the easy ability to change point B to point C in flight. This is a classic case of incorporating features "because we can."

---

* See Norman's (5) discussion of VCR complexity.

At the same time, because of interface problems in some early GPS systems, the GPS certification requirements are being written to require a complete ATC database before the system can be installed in an aircraft. This is simply not necessary.

Consider a typical light, single-engine aircraft. Most such aircraft contain a navigation system which can reference a few hundred ATC fixes — the VOR. To use this system, the pilot must enter an arbitrary string of numbers allocated to each fix with the chance for look-up errors, and is then able to navigate only directly to or from the fix (provided the fix is close to the aircraft). There is no means to store fixes and many ATC waypoints are not available. The time and distance to the fix are likewise not available. Yet with this crude system (and no additional navigation backup), a pilot can operate within the ATC system and make non-precision approaches down to 250 ft MDA and 1/2 mile visibility.(*12*) Further, an aircraft can operate in instrument conditions with a single such receiver.

Any aircraft GPS receiver (or a handheld receiver) can do this and much more. The pilot need only enter the identifier (or coordinates) for the fix; the system provides time/speed/distance information to the fix. The simplest of GPS receivers is easier to use and provides much more information than the best of existing navigation receivers.

Yet a GPS receiver is required to have the capability to store all ATC waypoints, store multiple waypoint flight plans, even for a GPS system restricted to visual flight conditions!(*13*) No one can argue the increased capability is valuable, yet the authorities have barred a simple GPS which would reduce pilot workload and increase capability.

This has come about, in our opinion, for two reasons. First, many of the early designs were poor designs with poor MMI. The certification requirements fix the symptoms — not the underlying cause. Second, we insist on the advanced features because they are available (i. e. because we can).

Restatement of the Problem

The underlying causes are the absence of a logical, organized design methodology with well defined requirements, user community feedback, and test and evaluation criteria.

The problem isn't complexity or lack of standardization. The problem is bad design practices.

3   A Review of Cockpit Design Guides

What is needed is a design methodology handbook that replaces the two present design approaches to display development: TLAR* or a slavish adherence to a standard. It is essential that a rational and effective design procedure be prepared.

What has happened in the past was summarized by Shafer:(*14*)

> A strategy often used by project managers is to assign ex-users (pilots, electronic warfare officers, tank drivers, etc.) to design the human interface. What results are two kinds of problems that cost a lot of money later in the system life cycle.
>
> First, the ex-user is often a "super-user" who performs well above average and assumes that future users will do the same. Second, ex-users, not usually trained in human factors/ergonomics, fail to use the proven design tools that ferret out interface problems early and produce good designs. A subtle third effect is that ex-users have the war story arsenal to defend their design decisions during those intense design reviews, with the result that human interface problems and subsequent design changes occur long after field test and evaluation. A costlier solution indeed. Shafer (*14*)

Previous Design Documents

The development of any display must start with the basic principle of analyzing the mission requirements. The information required by the pilot and crew must be cataloged. Only then can the display symbology be designed. Head-down instruments did not change greatly for many years. As a result, designers forgot this basic principle and concentrated on matching the format of the "basic T."

Ketchel and Jenney (*15*) reviewed the informational requirements of electronic displays. They outlined the general need for an informational requirements study and reviewed sixteen such studies. They charted the

---

* TLAR = That looks about right.

information requirements for each study and summarized them for selected phases of flight (takeoff, enroute, and landing). In their review, the needs of the pilot were assumed to be proportional to the number of times in each data item was mentioned — a vote basis. Jenney and Ketchel do mention that such a summation is no substitute for a detailed analysis, but only as an approximation of the needs.

As an example, Jenney and Ketchel mention a pull-up warning to avoid terrain. This was only listed twice (out of sixteen reports), but is obviously an important information item. This points out a major limitation of pilot surveys or summaries in determining informational requirements and the need for careful consideration of all relevant issues.*

Singleton (16) described a more generic approach to display design. The display design must consider why the pilot needs the data and what the pilot is expected to do with the data. According to Singleton, several questions must be answered during the development of a display:

➤ Does the pilot's need justify the display?
➤ What data does the pilot need that has not been provided?
➤ Can the average pilot obtain what is required easily?
➤ Does the display conform to the real world?
    To other cockpit displays?
    With previous pilot habits and skills?
    With required decisions and actions?

This last series of questions concerning conformity should not be taken as an absolute requirement for duplicating previous displays or the real world. Rather, it means that the display should not be in conflict with the pilot's experience and training nor with the external cues. It would be foolish to insist that HUDs and HMDs conform exactly to early round-dial instruments or electronic head-down displays.

Abbott (17) described a task-oriented approach in which the information related to a particular task was analyzed. He applied this to an engine display by describing the pilot's task in terms of controlling thrust, verifying the engine is operating correctly, and indicating a minor thrust

---

* Jenney and Ketchel mentioned sideslip information and concluded that it was of limited importance to fixed wing aircraft. This may reflect a large proportion of fighter aircraft in their survey sample. It may also reflect no thought for engine-out control.

deficiency. He tested the display and found both a pilot preference and in-creased error detection capability.

Billings (8) prepared a design document to aid the cockpit designer to develop human-centered automation. His conclusions are that

+ Automation should provide a range of options;
+ Automation should provide better and more timely information;
+ Automation should provide explanation of its actions and inten-tions;
+ Automation should monitor trends and provide decision support; and
+ Designers should provide simpler, more intuitive automation.

Sexton (18) describes a design methodology for cockpits/flight sta-tions. According to Sexton, the design team should contain a pilot who is intimately familiar with the mission and should remain intact from mission analysis through test and evaluation. He places heavy emphasis on devel-oping mission scenarios. Sexton's design methodology is shown in Figure 3.1. Note that there is no explicit feedback.

Palmer, Rogers, Press, Latorella, and Abbott (19) developed a flight deck design philosophy for the high speed civil transport (HSCT) airplane. This was primarily driven by the desire to eliminate transparent windshields and replace them with electronic sensors. The need to manage configuration changes during flight (subsonic/supersonic cg shifts) and adjust the trajectory to avoid sonic booms in certain areas also drove the desire for such a philosophy. They consider the many aspects of pilots as: Team members (allocation of tasks); Commanders (decision makers); In-dividual operators (psychomotor activities); and Flight deck occupants (en-vironmental and crashworthiness considerations). Figure 3.2 shows the process for the HSCT design.

Wilkins (20) considered flight deck/crew systems design and inte-gration for the short haul civil transport (SHCT) or civil tilt-rotor (CTR). He stated that one must consider the mission requirements (the need to use narrow, obstacle rich corridors) , unique aerodynamic characteristics and the desired flight profiles. The CTR must also deal with the requirements of the transition from helicopter to airplane modes and back again. Finally, Wilkins states that the cockpit must consider the career origins of the flight crew (i. e., will they come from helicopter or fixed-wing pilot communities). Wilkins recommends making use of existing, proven concepts and designs and surveyed the current state of the art in cockpits.

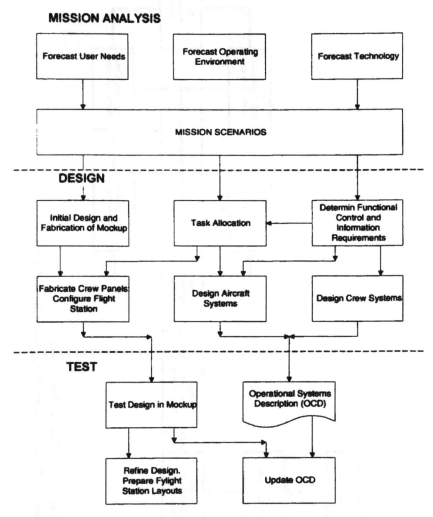

Figure 3.1   Design Flow Chart
Source: Sexton (18)

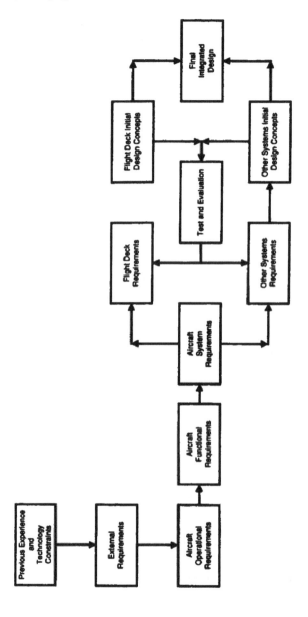

Figure 3.2   Design Flow Chart
Source: Palmer, Rogers, Press, Latorella, and Abbott (19)

Buchroeder and Kocian (*21*) reviewed the design trade-offs for a helmet-mounted display for the Army's Light Attack Helicopter. The study concentrated on the optical and physical integration issues.

Newman (*22*) prepared a HUD design handbook which was the result of two Air Force sponsored HUD studies to develop generic specifications for head-up displays. The study concentrated on symbology and systems integration issues and drew heavily on lessons learned from past programs. Newman also recommended a detailed informational studies (modeled after Singleton) and called for a logical test and evaluation protocol which was adapted from Haworth and Newman.(*23*)

Newman and Greeley (*24*) completed a HMD design handbook based on a review of HMD symbol and stabilization concepts. This handbook contained a HMD symbology database. Like the HUD handbook, it recommended a formal evaluation procedure.

Rogers and Myers (*25*) have developed an expert system approach to display design. This system, ACIDTEST, is designed to provide support for the display designer. The system provides guidelines to the designer to ensure all informational requirements have been considered. It also lists display rules and guidelines. Where conflicts exist, the system identifies these to the designer. Although promising, ACIDTEST has not been used in an actual systems design at this writing.

Storey, Rountree, Kulwicki, and Cohen (*26*) describe the Crew-Centered Design Process, developed at Armstrong Laboratory. This process has five steps: Planning, Requirements/Predesign, Crew System Analysis, Design, and Evaluation with feedback to previous steps. Their design flow diagram is shown in Figure 3.3.

Rasmussen (*27*) describes different levels of functional abstraction using means-ends relationships. This system, ecological interface design (EID), presents several levels: Purpose, priorities, work functions, specific work processes, and physical configuration. Intentional constraints flow down and physical constraints flow up. Rasmussen further divides operator functions as skill-based behavior, rule-based behavior, and knowledge-based behavior. Skill-based behavior comprises such task as steering or vehicle control. Rule-based behavior, such tasks as "remain to the right." Knowledge-based behavior includes cognitive decision tasks such as flight planning, diversion decision, etc.

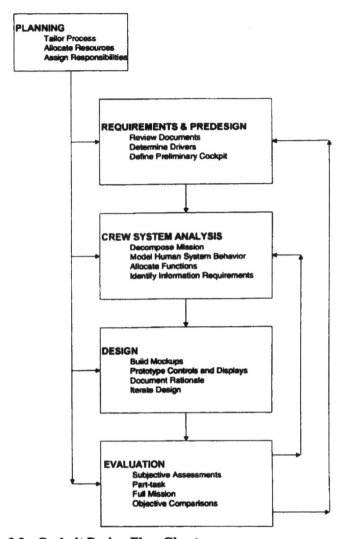

Figure 3.3   Cockpit Design Flow Chart
Source: Storey, Rountree, Kulwicki, and Cohen (26)

The Society of Automotive Engineers has prepared an Aerospace Recommended Practice (ARP) (28) which states that automation designers should Perform a detailed task analysis, Identify quantitative performance objectives, and Define the information requirements. Quantitative performance objectives are important for two reasons: their identification increases the likelihood that the proper tasks have been recognized and they will be used as criteria to measure the success of the design. The ARP also presents a design flow chart (See Figure 3.4).

Why Don't Design Guides Work?

These design documents all agree on the basic need for a detailed task analysis and information requirements study. Why don't they work? In our experience, it's because the designers generally don't really think that they need to perform such studies. The authors have worked on programs where, when tasked to develop display designs, were told not to perform a task analysis because "we did one once." Generally, the task analysis was never documented (or at least couldn't be found.

Cockpit design analyses are like system safety requirements. Everybody pays lip service — few actually perform the required preliminary studies. Norman (5) offers the explanation that the clients (in our case, procurement officers, contract administrators, or even chief pilots) are not the users (or operational crewmembers).

McDaniel (29) identifies the problem as the lack of a focus in the work breakdown structure in the military standards. He points out that there are ten level-2 elements in MIL-STD-881B.(30) Under the Air Vehicle element, there are seventeen level-3 elements. The crew system is spread among twelve of these elements. Without better focus, McDaniel suggests the problem will not improve.

There are other reasons, however. The first is cultural isolation. Each of the necessary disciplines, human factors engineers, electrical engineers, software designers, and flight crews operate in cultural vacuums.* None really understand what the others do. The technical personnel don't understand the mission (or at least aren't aware of recent developments and tactics). The flight crew aren't aware of the state-of-the-art in technology. This type of cultural isolation is often characterized by great attention to minutia, such as the microscopic dimensions of icons, and ignoring the

---

* Facilities layouts aggravate the problem. Personnel are segregated by task or are located in remote facilities, this acting as a barrier to face-to-face interchange within the design team.

global issues. Norman (5) describes this as "Designers are not typical users." Designers are more expert with the device; users are more concerned with the task.

A second reason for poor cockpit designs is the not-invented-here (NIH) mentality. We've always done it this way. Airplane B-model cockpits looks just like the A-model ones. Flight crew want the new cockpit to look just like the last airplane they flew. In other words, "we've always done it this way."

A common reason for many complex designs is "because we can." Designers fail to follow the KISS (Keep it simple, stupid) principle. Excessive data is presented on displays, many complex modes are loaded into FMSs because the technology allows it. This carries over to specification writing where the procurement or certification officials gold-plate the requirements.

The final reasons (perhaps the main reasons) are lack of operational feedback from the user community, combined with absence of design requirements.

Characteristics of Good Design Teams

What characteristics do good design teams have? First, they are composed of individuals with strong backgrounds in the various disciplines: human factors, electrical engineering, software. Generally the engineers have operational backgrounds, but recognize that their backgrounds may not be current.

Good design teams may come from joint ventures where no single company's engineers have the ultimate say. This may be a way to avoid the NIH problem.

The best design teams perform task and mission analysis using input from operational crews. They don't follow the inputs from the operational crews blindly, but temper it in the light of new technology, modified mission capabilities, etc. Most importantly, they test the design in part-task simulators initially, and later make good use of full-mission simulators. The results of these simulations are fed back and the design grows iteratively.

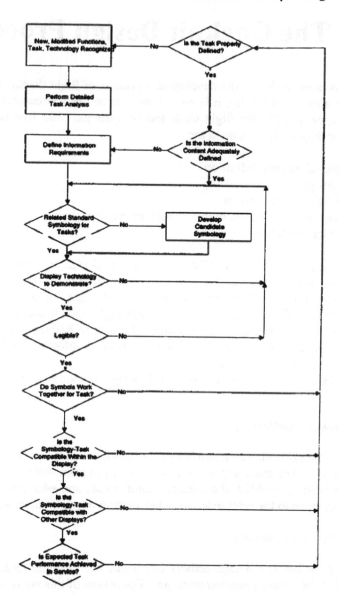

Figure 3.4   Cockpit Design Flow Chart
Source: *SAE ARP-5155 (28)*

4   The Cockpit Design Process

It is instructive to review the development process of flight display design. With the added complexity, it is more important to use a structured development approach for the flight deck and its systems. The five tasks required for display development are

- ✈ Requirements definition
- ✈ Design
- ✈ Engineering analysis
- ✈ Test and evaluation with feedback as required
- ✈ Documentation

Figure 4.1 shows how T&E meshes with design and development. The points to emphasize are the inclusion of a requirements definition task prior to the design itself, an early proof-of-concept T&E for novel designs, and feedback to the design process following flight and simulation testing. Feedback to the Mission Analysis and Information Requirements study may also be needed as a result of failure analysis or information deficiencies uncovered during testing.

Two survey papers outline cockpit design methodology.(*31-32*)

Requirements Definition

We have already reviewed a number of design guides in the previous chapter (page 12). The most important point to be made from that review is the absolute need to establish the mission requirements and what information can be provided to the pilot to assist in performing mission objectives.

Operational Requirements

Normally, before any design activity can begin, the customer must determine what the mission requirements are. The results of this study will be a requirements document, often referred to as the Operational Requirements Document or ORD. This ORD, or similar document, will form the basis for the customer's initial specifications.

The customer's procurement specifications should state what the display system must do and should be written in terms of objectives, performance requirements and should state the acceptance test methods. The specifications should define what acceptable test results will be. Clearly stating these requirements and acceptance criteria will prevent many of the divisive issues that arise during the development of new systems.

Mission Analysis and Information Requirements

Before any test and evaluation activities can take place — indeed before the design and finalized — a mission analysis and information requirement (MAIR) study should be performed and the results used during the design

The results of the MIAR study will also be used to develop the test plans for the display evaluation as well as assigning performance criteria for the display.

The MAIR study may performed by the customer organization (i. e., by the users) or by the design organization. If performed by the customer, it should form an adjunct to the procurement specification. If performed by the design staff, it should be a deliverable document. In either case, the final MAIR report should be explicitly agreed to by other customer and designers.

Recommended Approach

It is important to use the knowledge of subject matter experts (SMEs). However it is necessary that both operational pilots and experts familiar with the system under consideration be included. Recently Haworth and Lee at Ames Research Center have created a program to ensure that SMEs can determine if all required information has been ascertained.* Their approach asks pilots to indicate information requirements in normal and emergency flight. It also asks design engineers and others familiar with the system to indicate information requirements in their particular specialty.

---

*L. A. Haworth and A. G. Lee (Ames Research Center), to be published.

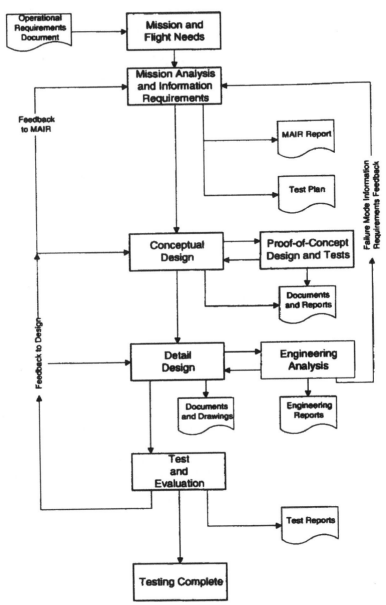

Figure 4.1   Design with Feedback from Testing

Haworth and Lee did well to include non-operational design engineers in their analysis technique. Often, novel systems have features that can impose mission requirements that SMEs not familiar with the particular system might miss.

Training Requirements

The MAIR study is similar to the training requirements studies described by Salas, Prince, Bowers, Stout, Oser, and Cannon-Bowers. (33). In their paper, step one was the identification of operational/mission requirements. This task was stated as

> Review existing training curriculum, including course master material lists, instructor guides, standard operating procedures (SOPs); interview aviation subject matter experts (SMEs); observe crews performing missions; review relevant mishap/accident reports.(33)

The results are specific to a particular mission context. While Salas *et al.* are concerned with improving existing CRM training, their approach can be extended to establishing the requirements for a new aircraft or a new display suite as well. Since training requirements and procedures will need to be established for any new design, it makes good sense to include training requirements while performing the MAIR study.

Design

In addition to the need for a formal, detailed mission analysis, many recent studies amplify the need for early test and evaluation feedback during the development of cockpit displays. These reports, discussed in the previous chapter, include Wilkins, (20) Storey *et al.*,(26) and SAE ARP-4155.(28)

Conceptual Design and Proof-of-Concept Testing

A significant problem with some systems being developed in recent years has been the deferring of any significant evaluation until the flight test phase. Often only a cursory simulator evaluation has been performed. On occasion, opinion polls of a paper design have supplanted any serious evaluation. The more successful programs, on the other hand, have incorporated early low-fidelity simulations in the early stages to obtain all-important feedback.

It is unrealistic to expect the designs to be right the first time. Changes made during flight tests are very expensive. Often, the high cost means that changes will not be made during flight testing, leading to design deficiencies in operational systems. Early proof-of-concept (POC) testing provides feedback, while changes are still relatively inexpensive, is essential. This requires a rapid prototyping system to make changes to the early display designs. Several very mature commercial systems* now exist for rapid prototyping and have been used on some recent fixed-wing aircraft development.

It the intended system is novel or has unusual characteristics, it may be necessary to perform early studies and develop preliminary designs for proof-of-concept (POC) testing. Such early POC testing allows for evaluations prior to commitment to a final design. This would reduce the overall program risk and allow changes early in the program when the associated costs are less.

POC testing should follow the general guidelines of this document, but can be performed on a less formal basis. Performance criteria for POC tests can likewise be less formal. Nevertheless, the results of any POC testing must be retained in a written document to support the lessons learned or design rationale reports. Even if the POC testing is sponsored in-house and funded separately from any procurement, the results should be retained and not lost to corporate memory.

Physical Constraints

Rolfe (34) describes constraints on cockpit design: Vehicle complexity, operational profile, and vehicle size and shape. These in turn affect flight deck in terms of contents, environment, and dimensions. The use of the human imposes psychological requirements for effective man-machine system performance, physiological requirements for survival and efficiency, and physical requirements for adequate workspace. He presented a cockpit assessment checklist (shown as Table 4.1).

---

* Such as Designer's Workbench® or VAPS®.

Table 4.1   Cockpit Design Checklist

Source: *Rolfe (34)*

1.0   General Information
1.1   Type of aircraft
1.2   Function
1.3   Number of crew members and their duties
1.4   Special clothing warn by crew
1.5   Arrangement of crew members in cockpit

2.0   Environmental Features
2.1   Entry and exit
    a   Normal: size and position of doors, access to seats, etc.
    b   Emergency: location, size, escape aids
2.2   Seating
    a   Position
    b   Alignment
    c   Shape
    d   Support (head, arm rests, etc.).
    e   Adjustment (of position, tilt, etc.).
    f   Strength (of construction, anchorage).
    g   Covering material
    h   Cushion material
    i   Restraint system
2.3   Stowage
    a   Maps and charts
    b   Reference books and checklists
    c   Aircrew clothing
    d   Food and drink utensils
    e   Emergency equipment
    f   Ashtrays
2.4   Lighting
    a   Cabin: intensity, variability, quality, reflections, etc.
    b   Panel: intensity, variability, reflections, alternative colours, effect on night vision, etc.
    c   Accessory: map reading, etc.
    d   Emergency lighting
2.5   Noise levels
2.6   Vibration
2.7   Thermal environment
    a   Cabin heating
    b   Air conditioning
    c   Connections for personal conditioning

Table 4.1   Cockpit Design Checklist (continued)
Source: Rolfe (34)

   2.0   <u>Environmental Features</u> (continued)
   2.8   Oxygen and pressurization
         a Normal
         b Emergency
   2.9   Survival

   3.0   <u>Information Input</u>
   3.1   External view (in relation to design eye datum point)
         a Angle of view: horizontal, vertical
         b Obstructions: pillars, etc.
         c Sources of glare: reflective surfaces, etc.
   3.2   Instrument displays
         a Layout (considered in relation to panel distances, impor-
            tance, frequency of use, sequence, function, and conven-
            tion
         b Individual design: readability, appropriateness, accuracy,
            failure indication, distinctiveness
   3.3   Auditory inputs
         a Intelligibility: quality of reproduction, speed of transmis-
            sion, number of possible inputs
         b Reproducibility: recorders, note pads
   3.4   Warning systems
         a Attention-getting ability: visibility, audibility, 'feelability' (e.
            g. stick-shaker); distinctiveness
         b Informativeness: indicating nature of trouble, advising
            remedial action

The Army and NASA have been working to develop a modeling tool, Man-machine Integration Design and Analysis System, MIDAS.(*35*) MIDAS is a simulation system incorporating cockpit dimensions, human anthropometry and vision models, with a human decision/memory model. MIDAS was developed to embed these various models and follow a simulation (not a piloted simulation) of various flight tasks. MIDAS is still under development. MIDAS is being used to develop the flight deck for the HSCT.(*36*). MIDAS appears to do electronically what present systems do with wooden mockups and task analyses.

Design Methods

Fruend and Sadosky (*37*) applied linear programming techniques to allocate instrument panel space. They concluded that the problem was complex!

Bartlett (*38*) illustrated the utility of using facilities allocation algorithms to display design. He developed alternative panel layouts based on eye-movement criteria. The paper did not state how the various mission tasks were weighted, but judging by the results, the preflight and engine run-up appear to have been weighted heavily.

Wickens and his co-workers (*39*) have developed the proximity compatibility principle (PCP) in which display elements used in related tasks should be close in location, color coding, or have connecting features. Andre (*40*) proposed a layout analysis. His conclusion was that display items should be grouped by flight task rather than by grouping frequently used displays near the center.

Dinadis and Vicente (*41*) applied the EID technique (*27*) to the engine and fuel systems display for a four engine turboprop using functional relationships.

Engineering Analysis

It is not possible to complete the validation process without conducting a number of engineering analyses. For aircraft displays, these include failure effects, reliability, fault tree and similar analyses.(*42-45*)*

The results will provide data to be used in test planning and will also be required in the final certification process. For example, Failure Modes and Effects Analysis (FMEA) describes the various failure effects. These results will be used to define the types of failures that will be used during testing to ensure that crewmembers can cope with systems failures. As shown in Figure 4.1, the analytical results may also loop back into the information requirements task to ensure that sufficient information is dis-

---

* Analyses to ensure that the installation meets structural airworthiness requirements, including crashworthiness are also required.

played to allow the crew to cope with system failures or other emergencies.*

Test and Evaluation

Following the development of the display design, a structured test and evaluation (T&E) process is required to continue the development (through the feedback to the designers) and to fine-tune the systems. While it is important that T&E be carefully thought out, often tests are carried out in a haphazard fashion or are poorly documented. T&E of any aircraft system has two purposes: Verification and validation.

Verification is the process of determining that a system's implementation accurately represents the developer's conceptual description and specifications. *Validation* is the process of determining the degree to which a system is an accurate representation of the real world from the perspective of the intended uses of the system. These definitions are taken from Army regulation 5-11.(*46*)

During the development process, the T&E effort seeks to determine how well the design meets its goals. Feedback from the test results will be used to improve the design to overcome shortcomings in the design.

In other words, the first purpose of T&E (verification) is to ensure that the system under test is the same as described in the various design documents, specifications, and system descriptions. The second purpose (validation) is to ensure that the system is useful in the real world and will meet the user's needs, contractual requirements, or minimum standards. This second purpose, in our opinion, is the ultimate reason for conducting test and evaluation.

Documentation and Certification

It is important that all design and evaluation decisions be documented. In particular, the design rationale for display design decisions, particularly controversial ones, be included as a formal document. This is becoming an increasingly important issue since there is becoming less continuity within

---

* Such as the requirement for annunciation of invalid data and standby displays to allow the pilot to cope with an instrument failure.

design and development organizations with the loss of corporate memory. Table 4.2 lists the documents developed during a typical program.

Even preliminary tests carried out in the initial portions of the T&E program should have the reports documented for the benefit of future designs.

Certification is the final step before operational use. Before the display can be certified for operational use, all required design documents, analytical reports, software test documentation, and test results must be prepared. The results should demonstrate a high degree of reliability and freedom from hazard. Depending on the results, appropriate operational limitations may be imposed.

Table 4.2   Documents Generated

Operational Requirements Document (ORD)
Mission Analysis and Information Requirements (MAIR) Report
Configuration Control Document (CCD)
Design Documents

> Crew Station Design Document (CSDD) [a]
> Drawings
> Specifications

Interface Control Documents (ICD)
Software Configuration Control Document
Software Release
Software Test Reports
Tabulation of Failure States
Test Plans
Structural Analysis of Installation
Electrical Power Analysis
Fault Hazard Analysis
Reliability and Safety Analyses
Training Syllabus
Ground Test Reports

> Environmental Test Report
> Electromagnetic Interference Test Report
> Optical Tests as required
> System Performance Tests

Simulator Test Reports [b]
Flight Test Reports [b]

Notes: a   Crew Station Design Document should contain design rationale for all symbols and icons.

b Test Reports should contain lessons learned during testing.

Certification is the formal process of documenting that the system meets the external formal requirements, whether contractual requirements or the minimum standards imposed by government regulators, such as the Federal Aviation Regulations (FARs). Certification determines that the system, including its documentation, analysis, and test results meet these externally imposed requirements.

Similarity

Is all this necessary for every new aircraft or modification to the cockpit displays? Installations that are similar to those previously approved should not be required to comply with all the steps described. The design analyses used for previous aircraft should suffice, provided the aircraft, displays, systems, and mission have not changed substantially.   This similarity, however, must be documented. Certainly, the reliability, loads, and structural studies will normally be completed for all designs.

Likewise, many of the specific test and evaluations may be omitted or reduced in scope on the basis of similarity. Functional testing should be conducted, at a minimum

However, if the aircraft, displays, systems, and mission have changed substantially, we would expect the design team to proceed with all applicable analyses and tests. The omission of any particular analysis or test on the basis of similarity should be justified and documented.

A few examples may help clarify this topic:

Change from "Steam Gauges" to Electronic Displays

For such a change, the MAIR study need not be duplicated, but the design team may rely on previous studies, if available. The full complement of structural, loads, electrical loads, and reliability analyses will be required. Assuming the display format is not too novel, minimal operational testing with line pilots should be needed. in any event, a complete unusual attitude recovery evaluation should be performed.

New Mission Task Added

For such a change, the MAIR study should be conducted. The design team may rely on previous studies, if available for those tasks that have not changed. The structural, loads, electrical loads, and reliability analyses need not be repeated if there is no change in the equipment. Operational testing for the new mission tasks must be performed. UA recovery evaluation can be omitted.

New Display Installed for New Mission Task

This might be the installation of a head-up display to allow category III landings. For such a change, the MAIR study should be conducted. The design team may rely on previous studies, if available for those tasks that have not changed. The structural, loads, electrical loads, and reliability analyses should be performed for the new display. Operational testing for the new mission tasks must be performed. For any new display, a complete UA recovery evaluation must be performed. In this case, however, the UA recovery study need only cover the use of the HUD during the recovery.

5   Requirements

This chapter summarizes several display standards and indicates their guidance for cockpit control/display test and evaluation. As will be seen, there is not much test and evaluation guidance from these documents.

Military Airworthiness Requirements

Many of these documents have been discontinued as part of the Department of Defense's initiative to reform the procurement process. This has left a number of programs without standardization documents.

Fixed-Wing Displays

MIL-STD-1787 (47) is the USAF standard for aircraft flight displays. While the standard covers all flight displays, the vast majority of the document deals with head-up displays based on fighter technology. Currently, there is no discussion of rotary-wing or VTOL aircraft, although these are planned to be included in the next revision.

The only discussions of test and evaluation call for visual inspection of the symbology (paragraph 5).

The intent is to include verification guidance in an appendix or supplemental handbook, there is little material in the current (Rev B) version. A draft for Rev C has been circulated with more guidance for non-fighter aircraft and additional guidance for test and evaluation.

The Air Force Flight Standards Agency (AFFSA) is responsible for ensuring that USAF aircraft meet the requirements of MIL-STD-1787 or have equivalent levels of safety. AFFSA outlines the flight test procedures. (48).A cross-section of pilots is to be used, including two pilots not familiar with the aircraft type in questions. Two of the pilots are to be operational pilots. All data is to be obtained via subjective questionnaires. Four flight test tasks are used in the endorsement process:

→ *Basic flight*. Takeoff/departure, navigation, instrument approach procedures, missed approach, and landing.

→ *Unusual attitude recognition and recovery.*

→ *Dynamic maneuvering.* Steep turns, vertical S-B and vertical S-D.

→ *Mission specific profiles.* These tasks will depend on the mission requirements and are intended for combat aircraft.

Rotary-Wing Displays

MIL-STD-1295,(*49*) was the first document creating a standard for helmet-mounted displays. It formed the basis for the AH-64, *Apache*, IHADSS helmet display. MIL-STD-1295 has been discontinued as part of procurement reform. There are plans to incorporate rotary-wing displays in MIL-STD-1787.

The Integrated Helmet and Display Sighting System (IHADSS) is the only operational helicopter HMD in service today. This is a monocular raster display with embedded symbols. All symbologies are screen-fixed. Figure 5.1 shows the symbology. The format appear to have been simply adapted from what would have been presented on a fixed HUD. Additional details can be found in the *HMD Design Guide.(24)*

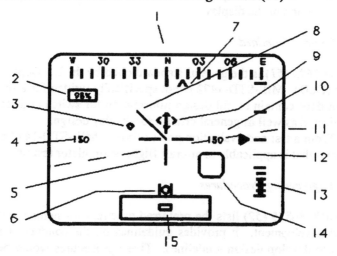

Figure 5.1   Standard HMD Symbology
Source: *MIL-STD-1295A (49)*

Data displayed in MIL-STD-1295A symbology (*Apache* HMD) include

[1] Heading	[8] Hover Vector
[2] Torque	[9] Head Tracker Reference
[3] Acceleration	[10] Radar Altitude
[4] Airspeed	[11] Non-Conformal Horizon
[5] Fixed Reticle	[12] Rate of Climb
[6] Sideslip	[13] Radar Altitude Tape
[7] Command heading	[14] Hover Position Box

MIL-STD-1295A does present an information requirements matrix as would be developed in a MIAR study. No test and evaluation criteria are discussed.

Fixed-Wing HUD

MIL-D-81641 (50) was the first (and to this point only) generic specification for head-up displays. It formed the basis for the original A-7 HUD. The document is written as an electronic procurement specification and outlines the procedures for qualification and production quality testing of the electronic devices. Only bench tests are covered — there is no discussion of validation of the display.

Human Factors Standard

MIL-STD-1472 (51) is the general standard cited by the US Department of Defense (DoD). MIL-STD-1472 is not specifically written for aircraft systems, but does contain useful design information for switch and secondary control design as well as general criteria for display design.

From a test and evaluation point-of-view, MIL-STD-1472 contains no material that is applicable to aircraft flight or simulator testing.

Human Engineering Procedures

MIL-HDBK-46855 (52) lists the preferred practices for human factors design and development. It provides guidelines on the conduct of mission analyses to develop design guidelines. These procedures would be useful in conducting a MAIR study.

From a test and evaluation point-of-view, MIL-HDBK-46855 describes procedures for conducting anthropometric evaluations. While the material describes such tests in cockpit mockups, it is applicable to ground form and fit tests.

Civil Airworthiness Requirements

US Airworthiness Requirements

FAR 23 (*10*) contains the US airworthiness requirements for light airplanes through commuter category. FAR 25 (*11*) contains the US airworthiness requirements for transport airplanes. The requirements for helicopters are contained in FAR 27 (*53*) and FAR 29 (*54*). The four regulations are written with similar requirements covered in similarly numbered paragraphs. Thus paragraphs 23.1309, 25.1309, 27.1309, and 29.1309 all deal with "Equipment, Systems, and Installation." Specific display and cockpit control paragraphs are listed in Table 5.1.

Systems and Equipment

Paragraph 23.1301 (and similar paragraphs in FAR 25, 27, and 29) requires the equipment to operate as intended. These paragraphs are used as the certification basis for systems for which there are no particular certification requirements. These paragraphs have been used for head-up display certification for both FAR 23 and 25 programs.

Paragraph 23.1309 (and similar paragraphs in FAR 25, 27, and 29) requires the equipment to operate without creating hazards, often referred to as *non-interference*. These paragraph require, among other things, the demonstration that equipment can pass environmental tests.

These paragraphs also indicate the levels of systems integrity based on the effect of failures. The effect of systems failures are:

Minor: Failure conditions that do not significantly reduce aircraft safety, and which involve crew actions that are well within their capabilities. Minor failure conditions may include slight reduction is safety margins or functional capabilities or a slight increase in crew workload. Minor failure conditions may be *probable*.

Table 5.1   FARs Applicable to Display Test and Evaluation

Paragraph Title	Paragraph Number			
	FAR 23	FAR 25	FAR 27	FAR 29
Pilot compartment	23.771	25.771	27.771	29.771
Pilot compartment view	23.773	25.773	27.773	29.773
Windshield & windows	23.775	25.775	27.775	29.775

Cockpit controls	23.777	25.777	27.777	29.777
Emergency evacuation	23.803	25.803	27.803	29.803
Equipment: function & installation	23.1301	25.1301	27.1301	29.1301
Flight & navigation instruments	23.1303	25.1303	27.1303	29.1303
Equipment, systems, & installation	23.1309	25.1309	27.1309	29.1309
Electronic display instrument systems	23.1311			
Instruments: arrangement & visibility	23.1321	25.1321	27.1321	29.1321
Airspeed indicating system	23.1323	25.1323	27.1323	29.1323
Static pressure systems	23.1325	25.1325	27.1325	29.1325
Instrument systems	23.1333	25.1333	27.1333	29.1333
Flight director systems	23.1335	25.1335	27.1335	29.1335
Instrument lights	23.1381	25.1381	27.1381	29.1381
Operating limitations & information	23.1501	25.1501	27.1501	29.1501
Airspeed indicator	23.1545	25.1545	27.1545	29.1545
Control markings	23.1555	25.1555	27.1555	29.1555
Airplane flight manual: general	23.1581	25.1581	27.1581	29.1581

Major: Failure conditions that reduce the capability of the airplane or the ability of the crew to cope with adverse operating conditions. Major failure conditions may include significant reduction is safety margins or functional capabilities or high crew workload or conditions that impair crew efficiency. Major failure conditions should be *improbable*.

Catastrophic: Failure conditions that prevent continued safe flight and landing. Catastrophic failure conditions should be *extremely improbable*.

Generally the following failure probabilities apply for critical systems failures.(*43, 55-57*) Recently, the failure rate requirements for light airplanes have been relaxed, based on the relatively high historical failure rates for vacuum-driven gyros and similar components.(*55, 58*)

Table 5.2   Probabilities for System Failures

Type of Aircraft	Maximum Probability		
	Minor Failures	Major Failures	Catastrophic Failures
Single piston engine, <6000 lb (*55*)	10^{-3}	10^{-4}	10^{-6}
Other airplanes < 6000 lb(*55*)	10^{-3}	10^{-5}	10^{-7}
Airplanes between 6000 and 12500 lbs(*55*)	10^{-3}	10^{-5}	10^{-8}

General rule for other airplanes(*43, 55-59*) 10^{-3} 10^{-5} 10^{-9}

Electronic Displays

AC-25-11 (*59*) provides guidance for the approval of electronic flight and navigation display systems for transport airplanes. Although somewhat out-of-data (the displays are limited to panel-mounted CRTs), the document has been used for other types of aircraft displays. The majority of the discussion concerns design issues and determination of criticality levels.

There is, however, some discussion of installation tests, including the requirement to move the attitude reference throughout all possible angles to demonstrate that the display can copy with extreme unusual attitudes.

AC-23.1311-1 (*58*) provides guidance for the approval of electronic displays for light airplanes. This advisory circular was recently updated to reflect the reconsideration of allowable failure probabilities. There is little material to guide test and evaluation.

The advisory circulars covering civil helicopter certifications (*56-57*) provide limited guidance for assessing flight displays. An example of such limited guidance is the recommendation that the standby attitude indicator be separated from the primary indicator:

> ...locating it [the standby attitude indicator] too close to the primary instruments may be undesirable and should be evaluated.(*60*)

This statement seems strange in view of the need to cross-check the primary attitude indicator with the standby. No explanation is offered and there is no further discussion is offered on how to conduct such evaluations.

Flight Test Guidance

The advisory circulars covering wing flight test (*61-62*) spend little space on evaluation of flight displays, other than calibrations of air data systems or electronic navigation sensors. AC-23-8 (*62*) has nine pages of flight calibration procedures for avionic systems. Specific autopilot failures are to be demonstrated.(*63*) The visibility of instrument failure flags is also discussed. However, there is no explicit discussion of evaluation of the displays for flight guidance. The advisory circular for transport airplanes presents a discussion of the handling qualities ratings, which will be discussed later (page 94).

Display flight test guidance for helicopters is likewise limited.(*56-57*) Because of the typical lack of inherent stability of helicopter, helicopters are restricted to visual conditions unless specifically approved. There is guidance that helicopters intended for IFR certification should be "flown in the air traffic control system in actual day and night instrument meteorological conditions."(*60*) This evaluation should include coping with failures, external visibility in precipitation, glare and reflections at night, workload, and handling qualities in turbulence, and cockpit leaks. However, there is no guidance on how to conduct these flights or in what data to collect.

Crew Workload

AC-25.1523-1 (*64*) discusses the establishment of the minimum flightcrew for transport airplanes. The advisory circular outlines a plan to demonstrate satisfactory workload beginning with analytical task studies. The final decision is deferred until the airplane has been flown by experienced and properly qualified pilots in representative operations. The flight data is based on structured subjective evaluations.

AC-25-1523-1 recommends that the test crews follow a representative duty day and the tests be dispatched with inoperative equipment to be permitted under the proposed minimum equipment list. For two-pilot aircraft, the tests should include the total incapacitation of one of the pilots.

The two helicopter AC's (*56-57*) are worded identically and recommend testing "during realistic operating conditions, including representative air traffic and weather." No guidance is provided on what data should be obtained.

The helicopter AC's discuss the difference between minimum crew required by the certification and that required by the operating rules. A particular aircraft may be flown by a single pilot in most operations, but will require two pilots others (such as during IFR flight or in commercial operations). They caution the manufacturer to consider all types of operations. It is implied, but not stated, that such aircraft be evaluated for acceptable workload with both single pilot and two pilot operations.*

Cockpit View, Reach, and Fit

Two AC's cover the design criteria and test methods for cockpit view.(*65-66*) These documents indicate that the external field-of-view requirement should be verified using a theodolite mounted at the cockpit design eye point (DEP).

AC-29-2C (*56*) goes into some discussion about conducting ground evaluations of fit-and-reach. The evaluation pilots are to adjust the seat to a flight position for the subject pilot. No mention is made of sitting with the pilot's head at the DEP. The evaluation should be tempered based on the anthropometric measurements of the particular pilot. Use of a number of different sized pilots is stated as an ideal.

Category II

AC-91-16 (*67*) covers category II approval for general aviation airplanes and requires an evaluation program of fifty approaches to demonstrate category II performance in flight. Of these approaches, ninety percent must be successful. A successful approach is defined as airspeed ±5 KIAS; glideslope ±50% of full scale; and tracking within the extended runway edges at minimums. For air carrier operations, three hundred approaches must be demonstrated with the substantially the same success rate.(*68*) Neither AC requires any data recording other than hand-recorded data.

Helicopter category II approach approval process requires the same numbers of approaches and success rate as for airplanes. However, test instrumentation must be used to record time histories of localizer and glideslope deviation and radar altitude.(*56*)

Category III

AC-120-28D (*69*) outlines the approval means for category III landing weather minima and low visibility takeoff minima. The performance requirements for category III (CAT 3) landing minima are quite rigorous.

---

* This is not normally tested in fixed-wing aircraft, although the King Air HUD described in Reference (*9*) did require evaluations in both single-pilot and pilot-copilot operations.

The requirements have recently been changed. Originally, there was a two-sigma requirement to keep the longitudinal dispersion less than 1500 ft and the lateral dispersion less than 27 ft,(70) but this requirement was omitted in AC-120-28D. The current requirements are

- Touchdown prior to 200 ft from threshold: $p < 10^{-6}$
- Touchdown beyond 2700 ft from threshold: $p < 10^{-6}$
- Touchdown more than 70 from centerline: $p < 10^{-6}$
- Exceeding structural limits at touchdown: $p < 10^{-6}$
- Exceeding bank angle limits at touchdown: $p < 10^{-7}$

These requirements must be demonstrated with a defined wind and turbulence model. They will normally require extensive simulations. Normally, these will be accomplished using non-pilot-in-the-loop simulation to establish the $10^{-6}/10^{-7}$ envelope. Sufficient piloted simulations should be flown to validate the simulator model. Finally, evaluation instrument approaches will be performed in flight. Normally, at least 1,000 simulated pilot-in-the-loop approaches and 100 actual aircraft landings will be required.

JAR Category II and III Requirements

The Joint Aviation Authorities (JAA) have prepared three documents covering requirements for evaluating HUD operations in all-weather operations.(71-73) These documents generally follow the requirements in FAA AC-120-28D (69) and 120-29.(68) However, very little guidance is provided for conducting the evaluations. JAR HUDS-902 does provide statistical guidance for establishing when the sample provides a ninety percent confidence of successful approaches.

Other Civil Requirements

The Society of Automotive Engineers (SAE) has published standards describing the performance requirements for head-down electronic displays (74) and for head-up displays. (75) They have also published recommended practices (76) for bench and optical testing for HUDs.

The Radio Technical Committee for Aviation (RTCA) has published the civil standard for environmental testing. (77)

Software Considerations

In equipment design, serious failures should occur less frequently than minor failures. As a result, the effect of a particular failure leads to the maximum failure rate of that failure. In software, failure rates have no meaning. As a result, software is design with various levels of integrity with the highest levels required for the more serious effects.

Military Software Criteria

DoD-STD-2167 (78) describes a formal process to develop mission-critical software systems for US DoD programs. The standard requires a detailed requirements specification, format test plans, and considerable documentation. Formal software testing must be conducted by individuals not involved in the coding. Several formal audits are required. At this time, DoD-STD-2167 has been canceled as a result of procurement reform.

Civil Software Criteria

RTCA-DO-178 (79) is the civil standard used for the approval of software. Revision B, dated December 1992 is the current standard, although Revision A, dated March 1985) is sometimes used.

DO-178A: DO-178A defines three levels of software: Levels 1, 2, and 3. Level 1 software, the most error-free is required for the most critical functions. DO-178A uses three criticality categories, based on failure or fault analysis.

> *critical*: Functions for which the occurrence of any failure condition of design error would prevent the continued safe flight and landing of the aircraft. Level 1 software is usually required for critical functions.
> *essential*: Functions for which the occurrence of any failure condition or design error would reduce the capability of the aircraft or the ability of the crew to cope with adverse operating conditions. Level 2 (or level 1) software is required for essential functions.

non-essential: Functions for which failures or design errors could not significantly degrade aircraft capability or crew operations. Level 3 software will normally suffice for non-essential functions.

DO-178B: DO-178B defines five levels of software: Levels A through E. Level A software, the most error-free is required for the most critical functions. DO-178B uses five criticality categories, based on failure or fault analysis.

catastrophic: Failure conditions that would prevent the continued safe flight and landing of the aircraft. Level A software is usually required for functions with catastrophic failures.

severe major: Failure conditions that would reduce the capability of the aircraft or the ability of the crew to cope with adverse operating conditions including large reductions in safety margins or functional capabilities, or higher workload such that the flight crew can not be relied upon to perform their tasks accurately or completely. Level B software is usually required for these functions.

major: Failure conditions that would reduce the capability of the aircraft or the ability of the crew to cope with adverse operating conditions including significant reductions in safety margins or functional capabilities, or significant increase in workload impairing crew efficiency. Level C software is usually required for such functions.

minor: Failure conditions that would not significantly reduce aircraft safety and involve crew actions well within their capabilities. Level D software is usually required for functions with minor effects.

no effect: Failure conditions that do not affect the operational capability of the aircraft or increase crew workload.

Software levels are usually related to the criticality categories, However, system architecture may allow for lower levels. For example, using multiple systems with dissimilar software, or separate monitors may allow a reduction in software levels. Instrument approach systems which normally require Level A have been approved with Level C software and a separate monitor using Level A software.

In addition, light airplanes, certified under FAR 23, may be developed with reduced levels commensurate with the relaxed requirements shown in Table 5.2. In other words, a system with catastrophic failure modes, normally requiring Level A software, may, in a light single-piston-engine airplane, be approved with Level C software.

Customer Specifications

Most of the requirements listed are across-the-board requirements which apply to all civil aircraft or to all military aircraft. Most procurements, other than commercial-off-the-shelf (COTS) will have some form of performance requirements stated as specifications. These can include weapon aiming requirements, aircraft performance, or may simply be stated as "Category II capable."

Performance Guarantees

During the testing for airworthiness releases, all airworthiness or safety issues will, of course, be addressed. However, an aircraft or its systems may be safe (i. e. airworthy) and, at the same time, fail to meet performance goals. During the conduct of these tests, performance measures will be obtained and a performance assessment prepared at or following the airworthiness release. With military aircraft, it may be difficult to separate performance from airworthiness and the same test and evaluation personnel will be conducting both tests.

With civil aircraft, it is usually easier to separate the two issues, the airworthiness review will be conducted by the national authorities (FAA, JAA, or other national CAA) and the performance assessment will be conducted by the airline or other customer.

On occasion, certification authorities have laid requirements on flight displays which, in the opinion of the customer, degraded performance. This has usually been the result of a lack of clear requirements, with the authority assuming one type of operation and the customer another. For example, a customer may desire a head-up display for use in austere landing zone operations. However, the authority may well include requirements appropriate for category III operations and entirely inappropriate for the intended use of the HUD. There is no easy answer, but a clear statement of the intended use will go a long way to help with this type of problem.

Human Factors Issues

Similar problems arise with human factors issues. Often human factors requirements are so nebulous as to be meaningless, or at best subject to the whim of the evaluating pilot. Having clearly stated requirements and clear stated *objective* test criteria will certainly help alleviate this type of issue. We also feel that confining the final test and evaluation (Mission Testing) to operational pilots will play a major role in preventing such problems.

6 Test Sequence Leading to Airworthiness

The clearance sequence for approving the airworthiness of the display systems should follow a logical progression. This requires completing a series of analyses, software testing, and preliminary ground tests before approving the display for flight. The display flight tests must follow a series of gradual expansion of tasks and environmental conditions to minimize risk. The overall progression is shown in Figure 6.1.

+ Supporting Analyses
+ Software Releases
+ Ground Tests
+ Simulator Tests
+ Preflight Release
+ Flight Test
+ Final Airworthiness Release

The various test sequences will be covered below.

Supporting Analyses

The various reliability and safety analyses, fault tree analysis, reliability analyses, as well as FMEA, will be used to ensure that the display system and its supporting sensors are sufficiently reliable to allow their operational use. A preliminary report showing an acceptable hazard level will be required for the pre-flight safety review.

Software Releases

The software should be developed following either the military standard, DoD-STD-2167 (78) or the civil standard, RTCA-DO-178.(79)

These two standards require extensive software testing before the final release. It is not necessary and probably counterproductive, for the ground and flight test software to be completely tested before flight testing. There will certainly be changes to the software following testing, both to remove special test routines and because of problems discovered during testing. For this reason, it is usually more practical to defer complete software testing until late in the test cycle.

Ground Test Software

Ground test software may have routines included to allow for data collection. In addition, systems failures may be simulated by other routines. The level of testing for the software used during ground and flight tests should ensure that it [the software] is fully functional in the areas under test. Reliability testing is not a pre-requisite for ground testing nor for simulator testing.

Flight Test Software

Flight test software requires further testing to ensure that it is free from hazardous effects. It is not necessary to have a complete software release for non-critical software functions for flight test.

The preflight software release can be split into pre-simulation and pre-flight releases if desired. The pre-simulator software release need not be held to the same no-hazardous effects requirement as the pre-flight release.

Final Software Release

Final software release requires complete testing to demonstrate as acceptable level of safety as defined in the software standards.(78-79) The test routines that inject errors into the data stream to simulate failures *must be removed* prior to the final release. Data collection routines may be left in the software.

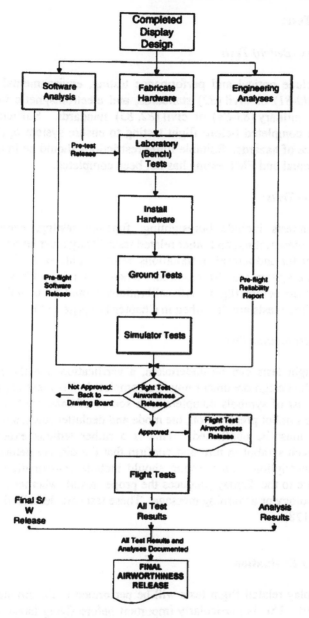

Figure 6.1   Display Airworthiness Progression

Ground Tests

Laboratory (Bench) Tests

These include optical and performance testing, environmental testing to military (80-81) or civil (82) standards, and electromagnetic interference testing to military (83-84) or civil (82, 85) standards. Sufficient testing should be completed before flight testing to ensure system operation and the absence of hazards. Suitable flight restrictions should be in place if the environmental and EMI testing has not been completed.

Installation Tests

Installation tests include boresighting, function testing, egress testing, reach and view testing, and other related tests. They must be performed for both simulator and aircraft installations before flight. In addition, calibrations of the system are also required. Where possible, calibrations should be performed before flight. All calibrations should be completed before testing. These tests are described in Chapter 10 (page 114).

Design Verification Test

Before flight tests can be undertaken, a verification that the installation matches the design document must be completed. This test requires that a complete list of symbols complete with source of data, control laws governing the symbol presentation, the mode and declutter rules, and occlusion priorities must be confirmed. This is a rather tedious exercise to go through each symbol in turn and confirm that the display behaves according the description. This testing should include confirmation that input data failure to the display produces the proper result, whether it is alternative symbology or a warning message. These tests are described in Chapter 10 (page 128).

Simulator Evaluations

Most display related flight tests will be performed in the simulator before the aircraft. This is particularly important before flying hazardous flights (such as unusual attitude recoveries) or tasks involving high workload (such as full mission simulations). The results of the simulations will be reviewed prior to clearing the aircraft for flight testing the same scenarios. Additional discussion of simulation evaluation is presented on page 86.

Pre-Flight Release

The pre-flight release is similar to final certification. Before the display can be released for flight test, available design documents, analytical reports, software documentation, and test results must be reviewed. The results should indicate a safety level appropriate for flight testing.

Appropriate operational limitations, such as restriction to day-visual conditions, may be imposed. As test results accumulate, further releases allowing night or restricted visibility operations would be expected. Such relaxation of the operational limitations can be justified by the availability of a safety pilot with standard instrumentation not dependent on the system under test.

Flight Tests

Flight test includes both aggressive and precise mission task elements (MTEs) and mission-related tasks. The MTEs are stylized maneuvers designed to determine the ability of the display to deal with both precise and aggressive maneuvers. The mission-related tasks are more complex mission scenarios that evaluate the cognitive aspects of the mission.

Embedded in these tests are evaluations to measure the ability of the pilot to maintain situation awareness and detect failure.

Final Airworthiness Release

The final airworthiness release requires the review of all design documents, analytical reports, software documentation, and test results. The results must indicate a safety level appropriate for the class of aircraft.

It is quite feasible to issue incremental releases with appropriate operational limitations if all testing is not completed. Initial release for day-visual conditions only would allow flight training to begin before the final release. As test results accumulate, further releases allowing night or restricted visibility operations would be expected.

Mission performance is a contractual issue between the manufacturer and the customer. As such mission performance does not normally enter into the airworthiness release except as it affects operational safety. In practice, in military programs, mission performance may be reviewed at the same time as the final airworthiness release and the distinction may become blurred.

The final airworthiness release will consider what equipment is required in various types of operations. An example is a requirement for a radar altimeter at night or in IMC; with no such requirement for day-VMC operations. Airworthiness will also dictate dispatch, minimum equipment lists (MELs), and continued operational rules for inoperable equipment.

7 Situation Awareness

What is Situation Awareness?

What is Situation Awareness?

> If no one asks me, I know. If I were to explain it to a questioner, I do not know.*

That maintenance of situation awareness (SA) is critical to flight safety is a truism. While SA is, no doubt, required for pilots and crew members to make correct operational decisions, it has been difficult to find precise definitions for use in developing test criteria. Defining SA as "knowing what's happening" doesn't lead to objective test criteria.

It is important to keep our discussion of SA in an operational setting with the aim of developing valid test criteria. To this end, *Situation Awareness* (SA) may be defined as correct perception of the current operational environment, comprehension of the effect of the environment on vehicle status, and the projection of vehicle status to the future.†

The details will require more details of the specific vehicle, systems, and operating environment. Gawron, Weingarten, Hughes, and Adams.(88) include the following elements: (We have added some description of aircraft-specific details)

1   The Internal States of the System
  a   aircraft systems
  b   powerplant operation
2   The External States of the System
  a   aircraft position   ⎫   relative to an
  b   aircraft velocity   ⎬   inertial frame
  c   aircraft acceleration   ⎭   of reference

---

* This is a paraphrase of St. Augustine, "What, then, is time? If no one asks me, I know. If I want to explain it to a questioner, I do not know."(86)
† This definition was adapted from Jones and Endsley (87) who defined SA as "perception of the elements in the environment within a volume of time and space, the comprehension of their meaning, and the projection of their status in the near future."

53

3   The Relationship Between the System and its Environment
4   The Environment in Which the System Exists
     a air traffic
     b threats
     c weather systems
     d air traffic control.

What we are really interested in, from an operational point-of-view is what the effect is on operational decisions — what does the pilot do when confronted as circumstances develop. Does the pilot make correct decisions? Does he have the correct information to make these decisions? Besco (89) described the attributes of good pilots by asking what should they do. A similar approach should be our criteria for developing SA test criteria — are the decisions made by the pilot correct?

Loss of Situation Awareness

Situation awareness (SA) may be defined the correct perception of the environment, the current situation, and the projection of future status. Jones and Endsley (87) describe three levels of loss of situation awareness (LOSA):

➤ Level 1:   Fail to perceive information or misperception of information
➤ Level 2:   Improper integration or comprehension of information
➤ Level 3:   Incorrect projection of future actions of the system.

Clearly maintenance of SA is critical to flight safety. Unfortunately, during flight in instrument meteorological conditions (IMC), pilots frequently suffer from a loss of situation awareness (LOSA).

Loss of Geographic Awareness

Perhaps the most fundamental form of LOSA happens when the pilot becomes unaware of his geographical position. In the extreme form, this can be described as "lost." However, in a less extreme form, he may simply be unaware of the aircraft's proximity to terrain. This can lead to controlled flight Into terrain (CFIT) in which the pilot files the airplane into the ground under complete control. The classic example of such an accident was the flight of a Boeing 727 into Mount Weather while on final approach to Dulles Airport in December 1974.(90)

A loss in awareness of aircraft position relative to the track indicated by instrument flight procedures is also important. Coyle and Krolak (91) discuss the issues related to maintenance of geographical situation awareness. With conventional navigation displays (i. e. non-glass), this involves considerable cognitive workload. In addition to citing the usual array of LOSA accidents, they indicate that 30% of initial instrument rating flight checks are failed, due to loss of situational awareness, mostly in holding patterns and non-precision approaches.

Collision with Obstacle

The pilot may have good terrain awareness, but not know his location relative to obstacles. This type of accident is referred to as "collision with obstacle."

Incorrect Performance Assessment

Collisions with obstacles may also be a result of the pilot's inability to assess the aircraft's flight path relative to the obstacle. In these accidents, the pilot may be well aware of the presence of the obstacle (or terrain), but either believes the aircraft will clear it or be unaware of aircraft performance.

An incorrect flight path assessment can also lead to a landing short of the runway. This is often true when adverse conditions make such assessment difficult. Typical adverse conditions include darkness, restrictions to visibility, or precipitation. Night circling approaches are particularly vulnerable. The recent MD-80 accident at LaGuardia is an example of such an accident caused by incorrect flight path assessment.(92) In incorrect performance assessment accidents, the pilot is aware of the terrain or obstacle, but simply misinterprets the aircraft trajectory relative to the obstruction.

High Workload During Diversions

Loss of geographic awareness can lead to increased pilot workload during diversions or other flight plan changes. If the pilot is confused to his geographical location or is not clear where his destination or proposed route is, he may make an incorrect decision for his initial heading or altitude. This was evident in the Boeing 757 accident near Cali in late 1995.(*93*) In this accident, the crew accepted a revised clearance to a different runway than originally planned. The increased workload during the reprogramming of the flight management system led to an incorrect turn away from the fix, rather than toward the fix.

Systems Monitoring

It can be difficult in many aircraft for the pilots to monitor their systems status during high workload portions of the flight. In particular failures of the flight instruments themselves may lead to misinterpretation and subsequent loss of control. In a 1974 accident, the crew inadvertently left the pitot heat off. When the pitot tubes iced over during the climb, the airspeed indications became erroneously high. Unfortunately the crew failed to detect the discrepancy and ultimately stalled the airplane resulting in a fatal crash.(*94*), It can be very difficult to deal with failed or invalid flight instruments even if the pilot realizes that there is a problem.(*91*)

During the investigation of a 1992 crash of a prototype V-22 tiltrotor, deficiencies in presenting system failure data were identified.(*95*) These deficiencies had no bearing on this particular accident since the display of additional failure messages would not have contributed to the pilot's ability to either correctly analyze the failure or apply corrective action. Nevertheless, the final mishap report to recommend improvements to the crew alerting system on the V-22, One such recommendation was to present more detailed messages, such as replacing the generic *INVALID ENTRY* with the specific *FREQUENCY OUT OF RANGE*, when a pilot enters an incorrect frequency, or an appropriate message depending on the data element to be changed. These and similar changes have been incorporated into the engineering/manufacturing development (EMD) and subsequent V-22 cockpit management systems (CMS).*

---

* Ryan Wilkins (Boeing Helicopter), personal communications, 1999-2000.

Spatial Disorientation

Spatial disorientation (SDO) of pilots has been a persistent problem in aviation since the first flight into (and probable spiral dive out of) a cloud. It has long been known that pilots are unable to maintain straight and level flight without visual cues.(96) Benson (97) defines SDO as "an experience occurring inflight in which there is a defect in the aviator's perception of the attitude or position of his aircraft or where conflicting perceptions give rise to confusion or uncertainty."

Loss of Control

The problem, however, is not SDO; the problem is the unusual attitude (UA). While some may consider a UA as caused by SDO, most UAs are caused by inattention. Once a critical UA has developed, the pilot may become disoriented by the high angular rates and extreme attitudes.

Two recent UA incidents dramatize this. One involved a HUD-equipped fighter, the other a transport using conventional instruments, pointing out that UAs can happen to any pilot in any type of airplane. Extreme pitch and roll attitudes can develop very quickly.

In the first incident, a USAF F-15 was inadvertently placed in a 90° bank when the pilot became disoriented during overwater maneuvers. During the resultant UA, the nose sliced down to an extremely low attitude. The airplane rapidly reached an inverted pitch attitude of -70°, passing 5000 ft. at an airspeed of 650 knots. Fortunately, the pilot recovered, using a split-S, pulling over twelve g's. The HUD videotape of this incident(98) has been widely distributed in an effort to prevent future incidents by increasing pilot awareness of SDO.

In the second incident, a Boeing 747, flying on autopilot with autothrottles engaged, was placed in a UA when an outboard engine lost thrust in cruise. As the speed decayed, the autothrottle increased thrust on the remaining engines while the autopilot attempted to maintain a wings-level attitude. When the autopilot disconnected, possibly because of high control forces, the airplane rolled to an extreme UA, ultimately reaching an inverted nose down attitude of 57°. The airplane then descended out of control from FL410 to 11000 ft. in the clouds. Fortunately, the disoriented pilot recovered the airplane when visual conditions were regained, but pulled over 4 g's (in a 2.5 g airplane) during the recovery.(99)

Instrument Interpretation

Instrument flight is a difficult task completely at odds with normal maintenance of ground-based orientation. The forces acting on the pilot are not reliable orientation cues. As a result, the pilot is forced to rely on foveal visual cues provided by the flight instruments.

There are two modes of visual image processing. The focal mode is directed by attention and requires that the image be viewed with the fovea. The peripheral system* serves spatial awareness, self-orientation, self-motion, and gaze stability. In contrast to the focal system, the peripheral system operates almost without decrement down to near the absolute threshold for light detection and is nearly insensitive to blur. The peripheral system operates across the entire retina independent of gaze direction and functions subconsciously *without requiring voluntary attention.*

Current aircraft flight instruments do not provide peripheral cues, but use the foveal vision. This requires cognitive interpretation by the pilot. As a result, considerable training is required to acquire and maintain instrument flight skills. The cognitive workload competes with other mental tasks in the cockpit, such as decision making and systems management.

Excessive Workload

Instrument Interpretation

As discussed in the previous paragraph, the cognitive workload required to interpret the basic flight instruments is a major contributor to workload while operating under instrument meteorological conditions. This is caused by the non-intuitive displays used since the advent of instrument flight. The amount of initial and recurrent training is an indication of the workload.

In particular, the currency requirements are significant. A pilot can remain away from flying for an extended time and still be able to control the an airplane in up-and-away *visual* flight. It would be unlikely that

---

*Strictly speaking, this vision mode covers then entire retina and not just the peripheral vision. Properly, the term *ambient vision* should be used instead of peripheral vision. However, we will use peripheral vision for the remainder of the paper because it is more easily recognized.

an instrument pilot could fly safely *on instruments*, after a similar absence from flying.

Complex Procedures

Competing for the pilot's workload capacity is the complex instrument flight procedures in today's instrument flight rules (IFR) environment. The standard aircraft instrument suite does not lend itself to quick interpretation of aircraft trajectory relative to assigned flight paths.

Techniques for Measuring Situation Awareness

There are a number of techniques developed for measuring situation awareness. These have been reviewed by Gawron *et al.* (*88*) and Gawron.(*100*) Alphabetically, these are

- China Lake Situation Awareness (CLSA)
- Crew Situation Awareness (CSA)
- Situation Awareness Global Assessment Technique (SAGAT)
- Situation Awareness Probe (SAP)
- Situation Awareness Rating Technique (SART)
- Situation Awareness Subjective Workload Dominance (SA-SWORD)
- Situation Awareness Supervisory Rating Form (SASRF)
- Other Measures of Situation Awareness.

We shall discuss each in turn. Following these discussions, we will add discussions of performance based measures.

China Lake Situation Awareness

The China Lake Situation Awareness (CLSA) rating scale was developed for use during flight testing, particularly during repeated flight tests or demonstrations. It uses the five point scale shown in Table 7.1.(*101*) CLSA asks the pilot to rate his SA following a task. Unlike SAGAT (*vide infra*), CLSA is a brief single scale allowing for a quick single digit rating. This feature was important in its development for use in single-seat airplanes. The rating can be spoken and recorded on cockpit voice tapes (usually the HUD tape).

Table 7.1   China Lake Situation Awareness Scale

	Description	Content
1	Very Good	Full knowledge of aircraft energy state, tactical environment, mission; Full ability to anticipate and accommodate trends
2	Good	Full knowledge of aircraft energy state, tactical environment, mission; Partial ability to anticipate and accommodate trends
3	Adequate	Full knowledge of aircraft energy state, tactical environment, mission; Saturated ability to anticipate and accommodate trends; Some shedding of minor tasks
4	Poor	Fair knowledge of aircraft energy state, tactical environment, mission; Saturated ability to anticipate and accommodate trends; Shedding of all minor tasks as well as many not essential to flight safety or mission effectiveness
5	Very Poor	Minimal knowledge of aircraft energy state, tactical environment, mission; Oversaturated ability to anticipate and accommodate trends; Shedding of all tasks not absolutely essential to flight safety or mission effectiveness.

The advantages of CLSA are its face validity and ease of use. The subjective nature of CLSA is the main disadvantage. The abbreviated format is an advantage in terms of ease of use, but a disadvantage in terms of describing the SA in detail.

CLSA's use in a flight test organization leads to further disadvantages. These are the restriction to use by test pilots (as opposed to operational pilots), the small sample size, and the restriction to benign flight environment. This last is related to the use during flight test in single-seat aircraft. For flight safety reasons, extremely low SA scenarios will not be flown. Thus the CLSA has not been exposed to the full range of SA.

These disadvantages are a function of its use and are not inherent within CLSA.

At this time, CLSA has not been validated.

Crew Situation Awareness

Crew Situation Awareness (CSA) uses expert observers to rate crew coordination and identify and rate crew coordination and errors as minor, moderate, or major operational errors.(*102*) The observers then develop information transfer matrices identifying time and sources of prompts and responses. CSA is restricted to multicrew aircraft. It has been validated in simulation.

Situation Awareness Global Assessment Technique

Situation Awareness Global Assessment Technique (SAGAT) stops the test simulation at random times and asks the crew to state their assessment of the current situation. Their answers are compared with the correct answers. The comparison provides an objective measure of SA.(*103-105*) The main advantage of SAGAT is that it measures SA directly — what the pilot thinks the situation is and what it really is. Crew members were able to retain their knowledge for up to six minutes following freezing of the simulation.

The questions (items of interest) must be appropriate and must be developed prior to the experiment. The data to be collected will be the current perceived environment (distance and bearing to navigation aids or waypoints, distance and bearing to other aircraft, the current position when the simulator was frozen, etc.) Table 7.2 lists typical SAGAT questions.

These questions are intrusive enough to make it impossible to continue the test scenarios. This leads to the main disadvantage of SAGAT, the need to stop the simulation. While this allows detailed questions, it does add an artificiality to the test. It also precludes the use of SAGAT during flight trials.

SAGAT is very useful in providing early assessment of SA. It is not very efficient in conducting acceptance evaluations of flight displays.

Situation Awareness Probe

Situation Awareness Probe (SAP) solves the main disadvantage of SAGAT — the need to stop the test. In SAP, crewmembers are asked at intervals to answer SA related questions during testing. Typical questions are "How far to the next waypoint?" "Where is the nearest hostile aircraft?" ... These questions are similar to those posed by instructor pilots (IPs) during airmanship training.

Table 7.2   Typical SAGAT Questions
Source: Snow and Reising (106)

Level 1 Perception of the Elements in the Environment

What is your pitch?

What is your airspeed?

What is your altitude above the ground (AGL)?

What is your barometric altitude?

What is your current bank angle?

What is the next waypoint number?

Would you characterize the terrain 1 mile in front of you as "low and hilly" or "mountainous"?

Is there a missile threat to your aircraft?

What is the distance to your target?

What does your FPM currently intersect (i.e., sky, a hill, etc.)?.

Level 2 Comprehension of the Current Situation

What direction is the nearest terrain that is above your current altitude?

In what direction is the nearest navigational hazard?

Identify the nearest threat.

How much are you currently above or below your commanded airspeed?

How much are you currently above or below your commanded altitude?

How far left or right are you from your commanded path?

Level 3 Projection of Future Status

If you were to continue on your current flight path with no further stick input, estimate how long before you would impact terrain.

How far is it to the next waypoint (or target)?

Will your commanded flight path climb up in the next ten seconds?

Will your commanded flight path slope down in the next ten seconds?

What direction, if any, will your commanded flight path turn in the next ten seconds?

SAP is similar in approach to SAGAT, except that the questions must be short enough not to interfere with continuing the flight. The information will, of necessity, be more limited than with SAGAT.

Situation Awareness Rating Technique

The Situation Awareness Rating Technique (SART) (*107*) is a questionnaire technique that concentrates on measuring the crewmember's subjective rating in three areas:

> ✦ Demands on Attentional Resources
>> Instability of situation: Likelihood of situation changing suddenly
>> Complexity of situation: Degree of complication of situation
>> Variability of Situation: Number of variables changing in situation
> ✦ Supply of Attentional Resources
>> Arousal: Degree of readiness for activity
>> Spare Capacity: Amount of attention left to spare for new variables
>> Concentration: Degree to which thoughts bear on situation
>> Division: Amount of division of attention in situation
> ✦ Understanding of the Situation
>> Information Quantity: Amount of information received and understood
>> Information Quality: Degree of goodness of information gained
>> Familiarity: Degree of knowledge of situation

SART is a subjective measure who's greatest utility is the breakdown of the evaluation pilot's perception into specific model descriptors. At the same time, this is its biggest disadvantage.

Situation Awareness Subjective Workload Dominance

Situation Awareness Subjective Workload Dominance (SA-SWORD) was adapted from the SWORD technique (vide infra, page 70). It uses judgment matrices to assess SA. The data requires a rating scale listing all possible pairwise comparisons of the tasks to be perfumed.(*108*) For example, a question might ask the evaluation pilot for a comparison between Display A in IMC-Night conditions and Display B in VMC-Day conditions. The test engineer analyzes the rating using geometric means. The large amount of administrative overhead makes SA-SWORD less desirable for use in flight testing.

Situation Awareness Supervisory Rating Form

The Situation Awareness Supervisory Rating Form (SASRF) is a peer rating approach that asks operational pilots and supervisory pilots to rate other pilots in the same unit. Carretta, Perry, and Ree (*109*) used SASRF to evaluate the SA characteristics of F-15 pilots. The basis for SASRF is that squadron pilots know who does and who doesn't have the "picture." It is not clear how SASRF will apply to flight test. It is more suitable for personnel selection or training purposes.

Physiological Measures of Situation Awareness

Several physiological parameters (eye activity or EEG) have been suggested as indirect measures of SA.(*88*) These include blink amplitude and duration, scan pattern, pupil dilation, EEG, etc.) The measurements are more or less unobtrusive, although the equipment may be difficult to install in a cockpit. The biggest problem is the lack of validity, that is, there is no direct relationship between what is measured and SA.

Performance Measures of Situation Awareness

It is our opinion that the true test of SA is what effect it has on pilot decisions and resulting aircraft performance (i. e., what is the result after the pilot makes his decision). These SA evaluations are based on the discussion of SA problems discussed at the beginning of this Chapter. They can be summarized as follows:

Situation awareness: In these evaluations, the aircraft is flown in a simulated mission scenario and a "situation" created. The situation can be a systems failure, such as a navigation signal, a sensor, or an engine. The pilot should be have been prebriefed to make a verbal call when he recognizes the failure and perform appropriate corrective action.

The situation can also be a contrived mission environment, such as a simulated threat, flight plan rerouting, our diversion to an new destination.

The objective data obtained will include the recognition time, deviations from desired recovery actions, and deviations from flight path during the recovery. In all cases, the correct recovery action must be determined and compared with the pilot's actual recovery action. Anderson *et al.* (*9*) used this approach to perform an evaluation on a civil head-up display.

Attitude awareness: In most attitude awareness evaluations, the aircraft is placed in an unusual attitude and the pilot instructed to recovery to straight and level. This has been evaluated in a number of studies.(*9, 110-112*) The data is generally recorded as reaction time, number of initial control reversals, and, in some cases, recovery time and altitude lost. Generally, simulation is used to conduct the majority of such studies with limited in-flight recoveries used to ensure that there is no significant difference between simulator and airplane.

Newman (*113*) proposed an attitude awareness evaluation that would present a simulated air-to-air tracking task that would require the pilot to monitor his attitude and other flight parameters. A similar technique was used by Gallimore, Brannon, Patterson, Nalepka.(*114*) to induce unusual attitudes in a simulator.

Performance assessment: In these evaluations, the pilot is asked to state the performance achieved during a given task. This figure is compared with actual performance. This is similar to the SAGAT approach (*vide supra*). This was used by Anderson *et al.* (*9*) during instrument approach evaluations.

Comparison

Table 7.3 lists the comparisons between the various SA measures. Which technique should be used? Figure 7.1 shows a flow chart to make this decision. This was modified from a lecture by Gawron.(*115*)

Techniques for Measuring Workload

There are a number of techniques developed for measuring workload. These are

- ✈ Control Motion
- ✈ Secondary Task Performance
- ✈ Subjective Questionnaires
  - Cooper-Harper scale

- • Modified Cooper-Harper scale
- • Bedford workload scale
- • Subjective Workload Assessment Technique (SWAT)
- • Subjective Workload Dominance (SWORD)
- • Task Load Index (TLX)
→ Other measures of workload

We shall discuss each in turn. Following these discussions, we will add discussions of performance based measures.

Table 7.3 Situation Awareness Components

Component SA Test Method	Internal State	External State	System to Environment	Environment	Suitable for Flight Test
CLSA	✔	✔			Yes
CSA		✔		✔	Yes
SAGAT	✔	✔	✔	✔	No
SAP	✔	✔	✔	✔	Yes
SART	✔	✔			No
SA-SWORD	✔	✔			No
SASRF		✔	✔	✔	No
Physiological	✔				Yes
Performance		✔	✔	✔	Yes

Measuring the amount of control motion is a measure of manipulative workload. This commonly used during tracking tasks as a measure of the difficulty.(*116*) Difficulty in tracking is related to the number of processes the human operator must perform in relating control output to display input. To minimize tracking workload, the number of mental integrations and differentiations should be minimized.(*117*) The simplest tracking task requires a control output proportional to the error signal.

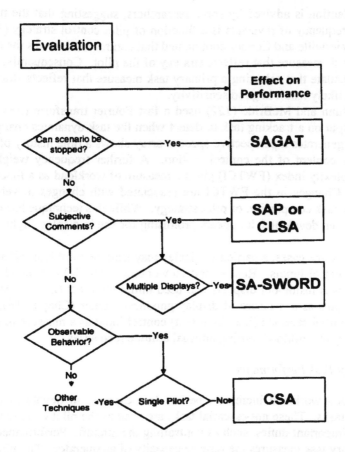

Figure 7.1   Flow Chart to Choose SA Measurement Technique
Source: Adapted from Gawron (118)

The number and frequency of control reversals has been used during flight testing as an indication of the workload caused by flying ILS approaches.(*119*) Since test aircraft control position is usually measured for stability and control purposes, such data is usually available. In a simulation task, Wierwille and Connor (*120*) found that control motion frequency was the only objective measure that demonstrated sensitivity to task load over their entire range.

Caution is advised by some researchers, suggesting that the number and frequency of reversals is a function of pilot control strategy.(*121*) Indeed, Wierwille and Connor commented that control motion was the only primary task measure that reflects strategy of the pilot. Consequently, one could speculate that selecting a primary task measure that reflects strategy will most likely result in good sensitivity.

Dunn and McBride (*122*) used a fast Fourier transform (FFT) of control input on a tracking task to detect when the task dynamics changed. The FFT generates a perspective spectral array shows a time history of the frequency content of the control motion. A further frequency weighted task complexity index (FWTCI) gives a measure of workload as a function of time. Changes in the FWTCI are associated with changes in vehicle dynamics, task complexity, or pilot strategy. While this technique has only recently been developed, it appears promising for use in display flight testing.

In some cases, a particular display may lead the pilot into initial erroneous control inputs. Roscoe and co-workers (*123-124*) examined outside-in versus inside-out displays and used initial incorrect (i. e., reversed) control inputs as a measure of display goodness. Later (Chapter 16), we shall use initial reversed (i. e. incorrect) control input as a measure of performance (not workload) during unusual attitude recoveries.

Secondary Task Performance

As a pilot's workload increases, he has less capacity to devote to non-essential tasks. These not-essential tasks are shed as the pilot concentrates on more important duties, such as controlling the aircraft. Performance on a secondary task measures the reserve capacity of an operator. The pilot is briefed to concentrate on the main task of aircraft control and devote only as much effort on the secondary task as the primary task will allow.

By measuring the performance on the secondary task, it is assumed we can infer the workload demands of the primary task.

The Workload Assessment Device (WAD) is a secondary task that was installed on the Calspan-operated NT-33 variable stability airplane.(*125*) It presented a letter or group of letters on the HUD. The pilot was tasked to respond with one button if the letter was in a group of assigned letters and with a second button if not. The performance is based on a combination of reaction time and percent of correct answers.

By varying the numbers of letters in the assigned group (in sizes of 1, 2, and 4 letters), the WAD allowed for determination of spare cognitive and perceptual workload. If the performance is plotted versus the number of letters in the assigned group, cognitive workload is proportional to the slope of the line. The intercept is a measure of perceptual workload.

In the previously cited study of simulated ILS approaches, Wierwille and Connor (120) found that time interval estimation was had sensitivity to some levels of task load. They found that digit shadowing, memory scanning, and mental arithmetic secondary tasks did not show significant measures of task workload.

Self-contained secondary tasks have also been used as a measure of workload. An example is the use of airspeed error as a measure of ILS tracking workload.(126)

Subjective Questionnaires

Subjective questionnaires simply ask the pilots how hard they are working on the assigned task. These questionnaires may be more or less structured or they may be free-form. The crew workload advisory circular (64) states that structured questionnaires should be used.

In general, the questionnaires address different aspects of workload. As in the case of situation awareness, the best approach to measuring workload, is evaluating whether or not the pilot has enough spare capacity to deal with contingencies. The evaluation team must design scenarios to stress the pilot (within the normal limits of the intended mission) and measure the resulting performance on tasks other than the primary control task. In other words, determine what the crew should do, and verify that they, in fact, do it.

Wierwille and Connor (120) found subjective ratings to match well with task loading while performing simulated ILS approaches and concluded that "well designed rating scales are among the best techniques for evaluating psychomotor load."

Cooper-Harper scale: The Cooper-Harper handling qualities rating (HQR) has been used to estimate subjective workload.(127-128) It must be emphasized that the HQ ratings were not intended as a workload measure but simply as a repeatable measure of how well the aircraft can accomplish a particular mission task (maneuver). For this reason, we do not feel that the standard Cooper-Harper handling qualities rating scale is a suitable tool to use as a subjective workload questionnaire.

Modified Cooper-Harper scale: Nevertheless, the walk-through logic of the Cooper-Harper scale is useful in developing other scales. One example is the Modified Cooper-Harper Scale developed by Wierville and Casali (*129*), shown in Figure 7.2.

Bedford Workload Scale: Another example is the Bedford Workload Scale (*101*) which is similar in logic, but with different decisions. It is shown in Figure 7.3.

Subjective Workload Assessment Technique: The Subjective Workload Assessment Technique (SWAT) was developed by the US Air Force's Aeromedical Research Laboratory and is based on workload having three components(*130*): Time Load, Mental Effort, and Psychological Stress.

   Each component is scaled from one to three. Prior to use, the evaluators rank-order the various combinations (twenty-seven in all). During the evaluation, each component is evaluated. Figure 7.4 shows a sample SWAT data card. Following the evaluation, the ratings are combined using the pre-evaluation rank ordering to provide a single numeric rating.

   The difficulty in using SWAT is the amount of pretraining and individual calibrations required for each evaluator. As such, SWAT may be inefficient for use in flight test. SWAT does have the advantage of developing ratings for the three separate workload components indicated.

Subjective Workload Dominance   Subjective Workload Dominance (SWORD) requires evaluation pilots to make a series of forced comparisons between tasks. In their original paper, Vidulich, Ward, and Schueren.(*131*) asked evaluators to compare workload during unusual attitude recoveries using one of six variations on a HUD. Fifteen pairwise comparisons were made for both minor and major unusual attitudes. Pilots were asked to complete the comparisons following a series of simulator recoveries. There was a correlation of 0.61 between reaction time and SWORD rating. Interestingly, using ratings from pilots who did not fly the simulator produced a correlation of 0.54! SWORD does not appear to be a useful flight test technique.

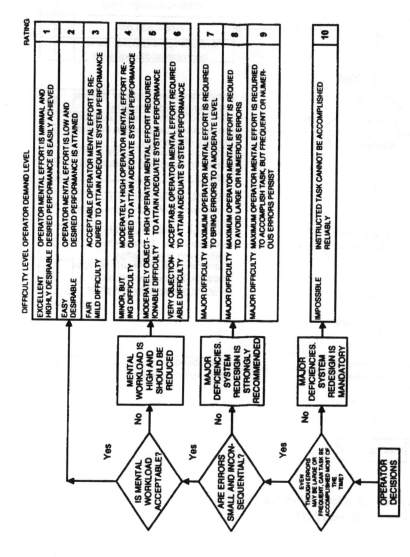

Figure 7-2 Modified Cooper-Harper Scale
Source: *Wierwille and Casali (129)*

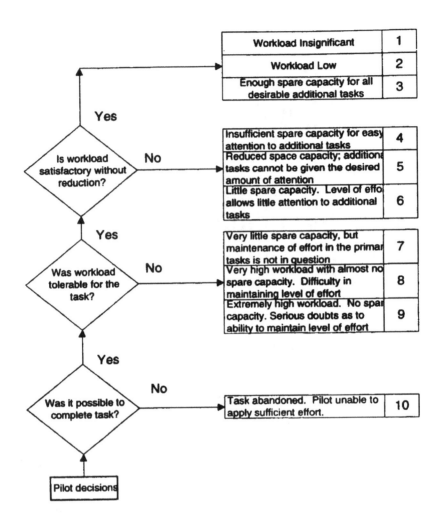

Figure 7.3   Bedford Workload Scale
Source: Adams (101)

Task Load Index: The NASA Task Load Index (TLX) workload scale
(*132*) is a structured questionnaires which generates a repeatable rating.

The questionnaires ask for ratings in six areas: Mental Demand, Physical Demand, Temporal Demand, Performance, Effort, and Frustration.

Evaluators are asked to rate each component as applicable to the task under question (in a pairwise fashion). This is used to determine the weighting for each workload component. Like SWAT, TLX has a significant amount of test overhead which may limit its usefulness in flight test. It is more suitable to simulator or laboratory trials. A computer tool is available.(*133*) Figure 7.5 shows a sample TLX data card.

Other Measures of Workload

Several physiological parameters have been suggested as indirect measures of workload.(*134*) These include blink amplitude and duration, scan pattern, pupil dilation, EEG, etc.) The measurements are more or less unobtrusive, although the equipment may be difficult to install in a cockpit. The biggest problem is the lack of validity, that is, there is no direct relationship between what is measured and workload.*

In the previously cited study of simulated ILS approaches, Wierwille and Connor (*120*) found that pulse rate had sensitivity to some levels of task load. They found that respiration rate, pupil diameter, voice patterns, and eye blink/eye transitions did not show significant measures of task workload.

Performance Measures of Workload

As with situation awareness, it is our opinion that the true test of workload is what effect it has on pilot decisions and resulting aircraft performance (i. e., what is the result after the pilot makes his decision). As a result, workload evaluations should follow along the lines of situation awareness testing discussed earlier in this Chapter.

---

* For example, there is little workload associated with riding a rollercoaster. Nevertheless, one's heart rate may be quite elevated.

SUBJECTIVE WORKLOAD ASSESSMENT TECHNIQUE	
NAME:	DATE AND TIME
TASK BEING RATED	

(MARK AND X ON ONE CHOICE FOR EACH OF THE THREE AREAS BELOW THAT BEST DESCRIBES WHAT YOU BELIEVE THE TASK WORKLOAD TO BE.)

I. TIME LOAD

[1] Often have spare time. Interruptions or overlap among activities occur infrequently or not at all

[2] Occasionally have spare time. Interruptions or overlap among activities occur frequently

[3] Almost never have spare time. Interruptions or overlap among activities are frequent or occur all the time

II. MENTAL EFFORT

[1] Very little conscious mental effort or concentration required. Activity is almost automatic requiring little or no attention.

[2] Moderate conscious mental effort or concentration required. Complexity of activity is moderately high due to uncertainty, unpredictability, or unfamiliarity. Considerable attention required.

[3] Extensive mental effort and concentration are necessary Very complex activity requiring total attention.

III. PSYCHOLOGICAL STRESS

[1] Little confusion, frustration, or anxiety exists and can be easily accommodated.

[2] Moderate stress due to confusion, frustration, or anxiety. Noticeably adds to workload. Significant compensation is required to maintain adequate performance

[3] High to very intense stress due to confusion, frustration or anxiety. High to extreme determination and self-control required.

Figure 7.4   Sample SWAT Data Card
Source: Charlton(128)

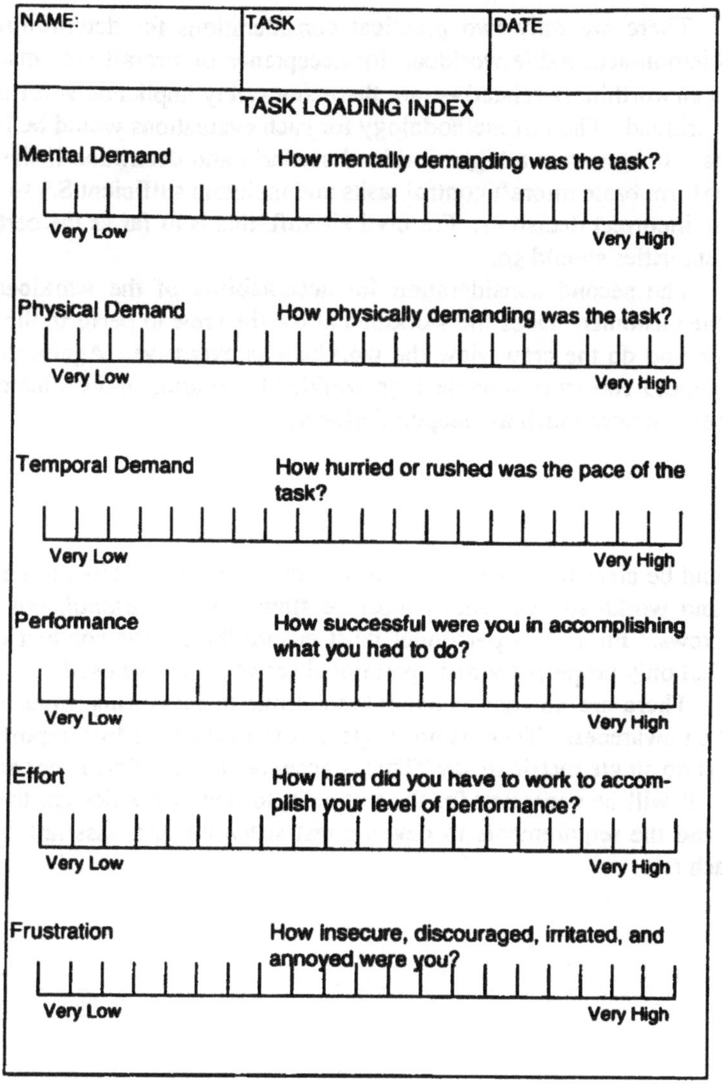

NAME:	TASK	DATE

TASK LOADING INDEX

Mental Demand   How mentally demanding was the task?

Very Low   Very High

Physical Demand   How physically demanding was the task?

Very Low   Very High

Temporal Demand   How hurried or rushed was the pace of the task?

Very Low   Very High

Performance   How successful were you in accomplishing what you had to do?

Very Low   Very High

Effort   How hard did you have to work to accomplish your level of performance?

Very Low   Very High

Frustration   How insecure, discouraged, irritated, and annoyed were you?

Very Low   Very High

Figure 7.5   Sample TLX Data Card
Source: Hart and Staveland(132)

There are only two practical considerations for determining acceptable/non-acceptable workload for acceptance of aircraft systems. The first is airworthiness-related — are there any safety implications for excessive workload. The test methodology for such evaluations would be for the test team to construct a high workload scenario and determine if the crew can perform basic aircraft control tasks and maintain sufficient SA to avoid making incorrect decisions. For civil aircraft, this is as far as the certification authorities should go.

The second consideration for acceptability of the workload lies with the customer. Does the workload allow the crew to perform the basic mission and do the crew view the workload as excessive. Again, the test team should construct suitable high-workload scenarios and evaluate mission performance (such as weapon delivery).

Summary

It should be clear from the discussions of this chapter that situation awareness and workload evaluations must be flown by *operational*, not flight test, crews. Flight test personnel must ensure the proper conduct of the tests, but only properly trained operational crews should be used

There are no easy cookbook solutions to measuring workload or situation awareness. There is no single task to evaluate. Most importantly, there is no single metric to say "This is acceptable" or "This is not acceptable." It will be necessary for the test team to review the design, the mission, and the requirements to develop test scenarios and pass/fail criteria for each test.

8   Flight Test

In this chapter, we shall discuss the test matrix and the sequencing of the testing to cover the required matrix.

+ Test Matrix
+ Evaluation Flight Tests
+ Visual Conditions
+ Simulation versus Flight Test
+ Test Sequence

Test Matrix

The test matrix for display evaluation has become more complicated with the advent of new display technology. Traditional round-dial instruments were generally only examined for adequacy of the information for instrument flight and the adequacy of night illumination. Electronic head-down displays require consideration of the color presentation in both low and high ambient lighting levels. Clutter, mode switching, and data latency became issues with the advent of electronic displays.

See-through displays, such as HUDs or HMDs, require that the background scene be considered as well. Head-mounted displays also are affected by pilot head-motion or head-tracking latency. As a result, the test matrix must now consider Evaluation tasks, External visual conditions, an pitot head motion. Figure 8.1 shows the conceptual complexity.

Tables 8.1 through 8.4 show how the complexity of the test matrix has increased for fixed displays.

As electronic displays were introduced, additional mission tasks in various lighting conditions were introduced (shown in Table 8.2). At about the same time, unusual attitude recovery evaluation became more common as a result of some operational incidents. Table 8.3 shows the increased test matrix for symbology-based HUDs. These were added because of difficulties with the full-scale motion of conformal HUDs which was complicated by excessive data latency in some aircraft.

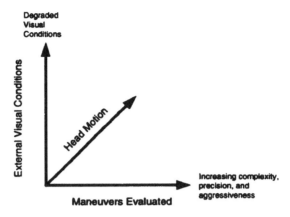

Figure 8.1   Complexity of Display Evaluation

Table 8.4 shows additional conditions required for raster image HUDs, to ensure satisfactory operation in degraded visual environments (DVE). Finally, as display evaluation progressed to helmet-mounted displays, the effect of head-motion had to be considered as shown in Figure 8.1.

Table 8.1 Test Matrix: Traditional Displays

External Visual Conditions	INSTALLATION	INITIAL TESTS	PRECISE TASKS	AGGRESSIVE TASKS	MISSION TASKS	SIT AWARENESS
		Test Evaluated				
VMC-day	■	■				
Solid-IMC					■	
VMC-night					■	
DVE-day						
DVE-night						

Table 8.2 Test Matrix: Glass Displays

External Visual Conditions	INSTALLATION	INITIAL TESTS	PRECISE TASKS	AGGRESSIVE TASKS	MISSION TASKS	SIT AWARENESS
		Test Evaluated				
VMC-day	■	■			■	■
Solid-IMC					■	
VMC-night		■			■	
DVE-day						
DVE-night						

Table 8.3 Test Matrix: Symbology HUDs

External Visual Conditions	Test Evaluated					
	INSTALLATION	INITIAL TESTS	PRECISE TASKS	AGGRESSIVE TASKS	MISSION TASKS	SIT AWARENESS
VMC-day	●	●	●	●		●
Solid-IMC			●	●		●
VMC-night		●				
DVE-day						
DVE-night						

Table 8.4 Test Matrix: Raster Image HUDs

External Visual Conditions	Test Evaluated					
	INSTALLATION	INITIAL TESTS	PRECISE TASKS	AGGRESSIVE TASKS	MISSION TASKS	SIT AWARENESS
VMC-day	●	●	●	●	●	●
Solid-IMC			●	●	●	●
VMC-night		●	●	●	●	
DVE-day			●	●	●	
DVE-night			●	●	●	

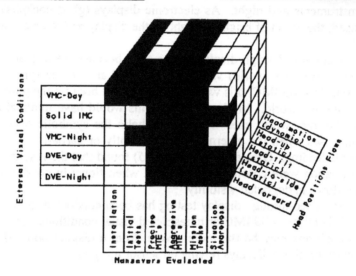

Figure 8.2   Test Matrix for Raster-Based HMDs

Evaluation Tasks

The use of both handling ratings and display ratings are based on flight tasks. These tasks should be related to the aircraft characteristics and the intended mission profiles. Mission task elements (MTEs) are simple tasks that represent mission tasks, such as precision hover, instrument tracking, or particular maneuvers. Display-related MTEs emphasize the dynamics of the display/aircraft, while HQ-related MTEs emphasize the dynamics of the aircraft itself. Some MTEs should stress precision while others should require aggressive maneuvers. Development of MTEs is discussed below.

Display evaluation also requires more complex tasks to evaluate the ability of the display to convey mission information. Mission-related-tasks are portions of a flight mission, such as the instrument approach (including terminal navigation) or ingress-weapon delivery-egress.

External Visual Conditions

Traditional display test and evaluation paid limited attention to external visual conditions. Evaluation of mechanical round dials only considered solid instruments and night. As electronic displays (glass cockpits) were introduced, the effect of bright sunlight on the display had to be considered as well.

Many see-through display issues arise because there is additional information (the real world) visible through and around the symbolic information. In solid IMC, this would have no effect because the pilot could not see the real world. In clear weather, this would have minimal effect because the real world cues would overwhelm the symbolic cues. The problem arises when the external cues are faintly or partially visible and may be misleading. Garman and Trang (135) found that the results of display evaluation were considerably different when flown in good visibility or when flown in degraded visibility.

Most see-through display testing has been conducted in either extreme of visibility, solid IMC or unrestricted visual conditions. The in-between conditions may be the more severe, from a display point-of-view, particularly for those displays relying on sensor imagery.

Pilot Head Motion

In the case of head-mounted displays, a final variable in the test matrix is the pilot's line-of-sight (LOS). Only with HMDs does the symbology/imagery remain visible when the pilot looks off-axis. The usefulness of the HMD must be evaluated as the pilot looks off-axis.(136) The effect of pilot head-tilt must also be examined to ensure that no difficulties arise from head tracker deficiencies.

Dynamic head-motion must be examined, with and without corresponding aircraft motion. For example, an aggressive turn to place an aiming symbol on a target initially 90 deg off axis is an aggressive MTE in ADS-33.(*137*) This should be modified for HMDs to require the pilot to turn his head to acquire the target, then turn the aircraft to point at the target.

Evaluation Tasks

The choice of evaluation tasks is dependent on the intended mission of the aircraft and on aircraft characteristics. The flight tasks are divided into mission task elements (MTEs) which are based on types of maneuvers likely to be flown during operations. These MTEs are intended to ensure that the display dynamics are satisfactory when flown in the aircraft. Both precise and aggressive MTEs should be flown.

Mission-related-tasks are based on flying full or part task portions of representative missions. The intent is to evaluate the cognitive workload of the crew as mission profiles are flown. Situation awareness tasks will be embedded in the mission-related-tasks.

Rotary-Wing Mission Task Elements

Table 8.5 lists MTEs which are appropriate for various rotary-wing missions. Some MTEs are obvious, such as turn-to-target for attack and scout aircraft. Some are chosen because of those display characteristics they will involve. For example, aggressive visual tracking is an excellent task for producing undesirable display interactions with the visual environment. For this reason, it should be performed on see-through displays intended for all missions.

Generally, the tasks in Table 8.5 were taken from the handling qualities tests in ADS-33.(*137*) Those ADS-33 tasks that are not particularly useful for demonstrating undesirable display tasks have been omitted, such as roll reversal, sidestep, and slope landings. These tasks are useful in separating good from bad FQs, but are not discriminating for displays.

Table 8.5   Mission Task Elements vs. Rotary-Wing Mission

Mission Task Element	Aircraft Mission						
	ATK	SCT	UTL	CAR	SOP	SAR	ASW
Aggressive MTEs							
Accelerations/Deceleration	X	X	X	X	X	X	X
Aggressive Visual Tracking	X	X	X	X	X	X	X
Bob-Up/Bob-Down	X	X	X		X	O	
Engine Failures	X	X	X	X	X	X	X
FCS Failures	X	X	X	X	X	X	X
Sidestep	X	X	X	X	X	X	X
Slalom	X	X	X		X	X	
Turn-to-Target	X	X	O		O		
Turn-to-Target, modified	X	X	O		O		
Unusual Attitude Recoveries	X	X	X	X	X	X	X
Vertical Remask	X	X	O		X	O	
Yo-yo, high	X	X					
Yo-yo, low	X	X					
Precise MTEs							
Hover	X	X	X	X	X	X	X
Hovering Turn	X	X	X	X	X	X	X
Instrument Maneuvers	X	X	X	X	X	X	X
Instrument Tracking	X	X	X	X	X	X	X
Landing	X	X	X	X	X	X	X
Pirouette	X	X	X	X	X	X	X
Mission Related Tasks							
Attack Profiles	X	X	O		X		X
Formation	X	X	O	O	X	O	O
Instrument Approaches	X	X	O	X	X	X	X
Terminal Operations	X	X	X	X	X	X	X
Training Assessment	X	X	X	X	X	X	X
Other Tasks as Required	O	O	O	O	O	O	O
Situation Awareness Tasks							
Failure Detection	X	X	X	X	X	X	X
Geographic Awareness	X	X	X	X	X	X	X
Attitude Awareness	X	X	X	X	X	X	X
Traffic or Threat Detection	X	X	X	X	X	X	X
Performance Assessment	X	X	X	X	X	X	X

Note: X means mandatory task; O means optional task
ATK=Attack; SCT=Scout; UTL=Utility; CAR=Cargo; SOP=Special operations; SAR=Search and rescue; ASW=Anti-submarine warfare

In addition, tasks have been added, such as a modified turn-to-target (helpful for evaluating rapid head motion for HMDs), aggressive visual tracking (helpful for evaluating poor display dynamics for HUDs and HMDs), and instrument tasks.

Details of proposed rotary-wing mission task elements are presented in Chapter 11 (page 152).

Instrument tasks have been included for all rotary-wing missions (optional for utility helicopters). We feel IFR qualifications is an essential requirement for all new aircraft. For aircraft to be restricted to Day-VMC only, instrument approaches may be omitted. The instrument related MTEs, instrument maneuvers and instrument tracking, should still be performed.

Fixed-Wing Mission Task Elements

Table 8.6 lists MTEs which are appropriate for several fixed-wing missions. These were chosen using the same approach as were used in the rotary-wing tasks using the FW handling qualities tests.(*138*) As in the RW tests, these MTE were chosen based on the display characteristics involved. Specific mission tasks, not included here, must be obtained from other sources, such as general instrument procedures.(*139-141*)

Details of proposed fixed-wing mission task elements are presented in Chapter 13 (page 185).

Mission-Related-Tasks

The purpose of mission-related-tasks is to increase the task complexity over the two previous paragraphs and thus increase the workload. Choice of specific mission scenarios will depend on the operational use of the aircraft. It is unlikely that all tasks for all aircraft can be listed in a single document. Specific mission tasks, not included here, must be obtained from other sources, such as the Aircrew Training Manuals (ATMs) (*142*) for other aircraft with similar missions. Performance criteria, required for DRRs and DFRs should be developed using the results of the mission analysis and information requirements (MAIR) study.

All mission-related-tasks should be specified by the customer. Instrument tasks should be flown for all aircraft, based on the presumption that all aircraft will be qualified for instrument flight. Details of sample mission-related-task are presented in Chapter 13 (page 209).

Table 8.6   Mission Task Elements vs. Fixed-Wing Mission

Mission Task Element	Aircraft Mission					
	ATK	FTR	REC	BMR	CAR	TKR
Aggressive MTEs						
Acrobatics	X	X	X			
Aggressive Visual Tracking	X	X	X	X	X	X
Air-to-Air Visual Tracking	O	X				
Engine Failures	O	O	O	X	X	X
FCS Failures	X	X	X	X	X	X
Formation	X	X	O	O		
Refueling	X	X	X	X	O	O
Steep Turn and Reversal				X	X	X
Turn-to-Target	X	O	O			
Turn-to-Target, modified	X	O	O			
Unusual Attitude Recovery	X	X	X	X	X	X
Yo-yo, high	O	X	O			
Yo-yo, low	O	X	O			
Precise MTEs						
Instrument Maneuvers	X	X	X	X	X	X
Instrument Tracking	X	X	X	X	X	X
Mission Related Tasks						
Attack Profile	X	X	X	X		
Formation				O	O	O
Instrument Approaches	X	X	X	X	X	X
Terminal Area Maneuvering	X	X	X	X	X	X
Training Assessment	X	X	X	X	X	X
Other Tasks as Required	O	O	O	O	O	O
Situation Awareness Tasks						
Failure Detection	X	X	X	X	X	X
Geographic Awareness	X	X	X	X	X	X
Attitude Awareness	X	X	X	X	X	X
Traffic or Threat Detection	X	X	X	X	X	X
Performance Assessment	X	X	X	X	X	X

Note: X means mandatory task; O means optional task
ATK=Attack; FTR=Fighter; REC=Recconaisance;
BMR=Bomber; CAR=Cargo or Transport; TKR=Tanker

Choice of Performance Objectives

The performance objectives shall be indicated in the procurement specification. They should be developed from the ORD. In some cases, these objectives may require modification following the MAIR study. Any such changes shall be approved by the procurement activity.

For tasks developed from aircrew training manuals (ATMs), the performance criteria in the ATM should be used as acceptable performance standard. Desired performance criteria should be somewhat better than the acceptable criteria.

Acceptable performance:	ATM performance criteria (i. e., track localizer within 1 dot)
Desired performance:	1/2 to 2/3 ATM performance criteria (i. e., track localizer within 1/2 dot)

Situation Awareness Testing

Situation awareness (SA) includes monitoring internal systems status and the external environment (such as monitoring for other traffic). Some of these tasks (such as coping with engine failures and monitoring for external threats or traffic) should be embedded in MTE and mission testing.

- Failure detection
- Geographic awareness
- Attitude awareness
- Traffic or threat detection
- Performance assessment

Failure detection: System failures should be introduced at random points during the MTE and mission-related-tasks. The pilot's reaction time and resulting performance should be measured.

Geographic awareness: During mission tasks, diversions from the planned scenario should be introduced. Typical diversions include threats, changes in flight plans, etc. Appropriate performance standards should be agreed to for these tests.

Performance assessment: During MTE and mission-related-tasks, the pilot should be briefed to report his performance. These results should compare his assessment with the actual performance.

SA testing will be embedded in the mission-related-task testing and is an essential feature of mission-related-testing. Representative situation awareness tests are presented in Chapter 14.

Choice of Visual Conditions

The choices of which visual conditions to perform the MTEs is dependent more on the nature of the task than the intended mission of the aircraft. Table 8.7 indicates visual conditions are appropriate for each rotary-wing MTEs. Some conditions are obvious, such only performing aggressive visual tracking in clear conditions. In any event, the specific visual conditions for which the display will be evaluated should be described in the procurement specifications.

Table 8.8 indicates visual conditions are appropriate for each fixed-wing MTEs. As stated above, the specific visual conditions for which the display will be evaluated should be described in the procurement specifications.

Simulation versus Flight Test

Simulation is used to reduce the risk involved in flight testing displays and other critical systems — both risk of mishaps and development risk. Simulation can also be used to supplement flight evaluations, thus reducing the amount of flight testing. While it is possible to conduct a display evaluation without simulation, this is not normally an effective approach.

It is important to remember that flight evaluation is the final adjudication for any display. The simulation should be designed to support the flight test process, either to reduce the risk or provide statistical samples.

Risk Reduction

The general rule for display evaluations is to test in the simulator first, particularly for novel displays or where the pilot-vehicle interaction is important. This enables practice to be gained without hazard and at a lower cost. All adverse environmental conditions should be flown first in the simulator, then in the aircraft.

Table 8.7 Rotary-Wing Mission Task Elements vs. Visual Conditions

Mission Task Element	Day	Night	DVE	IMC
Aggressive closed-loop tasks				
Accelerations/Deceleration	X	O	O	
Aggressive Visual Tracking	X			
Bob-Up/Bob-Down	X	X	X	
Engine Failures	X	X	X	X
Flight Control System Failures	X	X	X	X
Sidestep	X	O	O	
Slalom	X	O	O	
Turn-to-Target	X	O	O	
Turn-to-Target, modified	X	O	O	
Unusual Attitude Recovery				X
Vertical Remask	X	O	O	
Yo-yo, high	X			
Yo-yo, low	X			
Precise closed-loop tasks				
Hover	X	X	X	O
Hovering Turn	X	X	X	
Instrument Tasks				X
Instrument Tracking				X
Landing	X	X	O	
Pirouette	X	O	O	
Mission Related Tasks				
Attack Profiles	X	O	O	O
Formation	X	X	X	O
Instrument Approaches				X
Terminal Operations	X	X	X	X
Training Assessment	X	X	X	X
Other Tasks as Required	O	O	O	O
Situation Awareness Tasks				
Failure Detection	X	X	X	X
Geographic Awareness			X	X
Attitude Awareness			X	X
Traffic or Threat Detection	X	X	X	
Performance Assessment	X	X	X	

Note: X means mandatory condition; O means optional condition

Table 8.8 Fixed-Wing Mission Task Elements vs. Visual Conditions

Mission Task Element	Day	Night	DVE	IMC
Aggressive MTEs				
Acrobatics	X			X
Air-to-Air Visual Tracking	X			
Air-to-Ground Visual Tracking	X	X		
Engine Failures	X	X	X	X
Flight Control System Failures	X	X	X	X
Formation	X	X	O	
Refueling	X	X	O	
Steep Turn and Reversal	X	X	X	X
Turn-to-Target	X	X	X	
Turn-to-Target, Modified	X	X	X	
Unusual Attitude Recovery				X
Yo-yo, high	X			
Yo-yo, low	X			
Precise MTEs				
Instrument Maneuvers				X
Instrument Tracking				X
Mission Related Tasks				
Attack Profiles	X	X	O	X
Formation	X	X	O	O
Instrument Approaches				X
Terminal Operations	X	X	X	X
Training Assessment	X	X	X	X
Other Tasks as Required	O	O	O	O
Situation Awareness Tasks				
Failure Detection	X	X	X	X
Geographic Awareness	X	X	X	X
Attitude Awareness			X	X
Traffic or Threat Detection	X	X	X	
Performance Assessment	X	X	X	X

Note: X means mandatory condition; O means optional condition

Development risk reduction can be enhanced by flying the complex mission scenarios first in the simulator under well controlled conditions. This can usually be performed earlier in the program in the simulator than in flight. If problems occur requiring redesign, it will usually be cheaper to find out in simulation than late in flight test.

Simulation also allows full mission testing using complete crews of operational pilots which may not be possible in flight.

Statistical Samples

The simulator is also better suited for obtaining sufficient statistical samples. Many evaluation tasks require significant replications for validity. Flying these entirely in the aircraft can be time consuming. In these cases, it is well advised to use the simulator to obtain a sufficient sample size and perform limited flight trials. The flight trials would require only a sufficient sample size to demonstrate that the simulator population and the aircraft population were in agreement. Past programs have made use of this by performing evaluations with a large sample of evaluation pilots in a simulator. Following the simulator evaluation, a subset of these pilots were used in a flight evaluation. Development of the USAF standard HUD format followed this approach.(*111, 143*)

Perhaps the most significant use of simulation to provide statistical samples is in the certification of HUDs for category III landing aids. The requirements for landing performance, contained in AC-20-57A (*70*), is stated in a touchdown dispersion envelope which must contain Ninety-five percent of all landings. The normal means of compliance is to perform a simulation using a pilot control model to generate the statistical sample. Several thousand approaches are required. These simulations use a computer model of the pilot. Following these approaches, several hundred simulations are flown with an actual pilot. These simulations should confirm the computer model simulations. Finally, one or two hundred actual approaches in the airplane are flown. Again, these should demonstrate that the airplane performance matches the previous manned and computer model simulation results. Far fewer airplane approaches are required than would be necessary if all data were obtained in flight.

Tasks Unsuitable for Simulation

Certain evaluations are of questionable value if performed in a simulator. An example is aggressive visual tracking. This is a task that evaluates the display dynamics in conjunction with the airplane dynamics. Temporal variations in the response of the visual scene, the motion base, and the see-through display can inject artifacts into the data. Normally, this will not be a problem during precise flight or in instrument conditions. However the combination of aggressive pilot input and use of the simulated real world scene will present problems. Visual scene latencies of the order of 100 msec or longer should not be used. Even shorter latencies will be required for agile aircraft, such as fighter or attack aircraft.

A second type of evaluation for which simulation is not suitable involves visual accommodation issues. Many visual scenes in simulators are focused much closer than the virtual image display. Any study involving visual eye response, such as traffic detection of eye strain over long periods of time, should user simulated flight with great caution.

Tasks Unsuitable for Flight Test

Generally extremely hazardous evaluations, such as extreme unusual attitudes or critical workload situations should not be performed in flight. In particular, extreme unusual attitude recoveries and the spatial awareness task are more suitable in a simulator than in an aircraft.

Test Sequence

Prior to conducting flight evaluations, the display system should be examined to ensure that the system being tested matches the description in the various design documents and drawings. This is normally accomplished with a powered-up aircraft on the ground, although some inflight verifications may be required.

Any required inflight calibrations should be flown before performing the evaluation tasks. These include calibration of pitot/static system, calibration of hover vector (rotary-wing and VTOL only), and other calibrations. Obviously, these can not be performed prior to flight, but will be performed during initial flight trials.

The general approach for evaluating displays should be to progress from evaluation aircraft handling in good visual conditions (with no display) through use of the display in gradually degraded visual conditions as shown in Figure 8.3.

Configuration Control

It is a fact of flight test that the airplane changes from flight to flight. It is equally a fact of life that the test team must know exactly what it is they're testing. This problem is difficult enough when one black box is replaced with another or subtle changes in the external configuration is made, often to correct one or another problem. With all these changes, it is difficult to reconstruct what was really installed on the airplane unless careful records are kept.

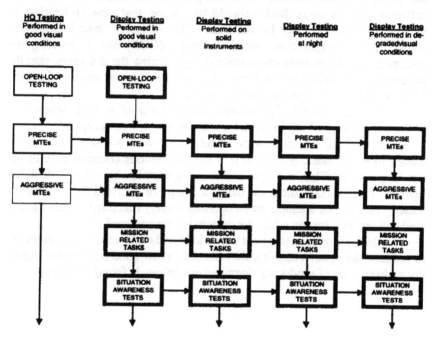

Figure 8.3   Flying Qualities Test Progression

If this problem is difficult with hardware changes, it is equally so with software changes. These changes can be made with no external change in the hardware. If fact, the data can be entered by keyboard with no outward notice. It must be possible for the test team to reconstruct exactly what was flown and when.(*144*)

One of the flight test horror stories was reported by Newman and Anderson, (145) who described a program where several software changes were made throughout a simulator and flight test program. Unfortunately, no record was kept of when these changes were made. The changes were substantive enough to change the display behavior significantly. The result: the tests, simulator and flight tests were worthless.*

At the risk of increasing the paperwork requirement, it is essential that the hardware *and* software configuration of every significant device on the airplane be documented. The test plan should clearly state what the software loads in every device should be. Some, that have no bearing on the test in question, may be optional. This however should be clearly stated. This configuration control document (CCD) will be an on-going document which contains the current design, listing the drawings, specifications, etc. installed on each test airplane. Following certification, the CCD should list all approved configurations.

A separate CCD for software should be used as well, although a single document for both hardware and software may be acceptable.

The flight documentation should state what software load is installed in every black box, even if it has no obvious connection with the test in question. This is simple insurance against having the test declared invalid. Furthermore, the crew needs a means of confirming software releases. This can be a simple as an opening display that displays the current load and the date loaded when the system is first powered up.

Because solid instrument conditions and night conditions are easier to produce than poor visibility, they should follow good conditions. Additional discussion of producing degraded conditions is found on page 104.

---

* This may be an overstatement. A case could be made that the first simulation run (with the original software load and the final flight test (with the final software load) could be acceptable.

9   Test Techniques for Flight Displays and Equipment

Rating Scales

The general approach to evaluate display qualities is similar to the approach of flying qualities (FQ) testing and handling qualities (HQ) testing. Flying qualities are the intrinsic characteristics of the aircraft and are independent of the visual conditions or the display system installed. They are normally determined in optimum visual conditions. Handling qualities (HQ) are related to these characteristics when the aircraft is flown in specific flight tasks in specified visual conditions. Thus, there is a natural progression from FQ to HQ in good visual conditions to HQ in degraded conditions.

The measure of success of any aircraft is how well does it perform in its intended task. The measure of success for an aircraft display is how well does it allow the pilot to control the airplane to perform the intended task.

HQ testing is flight task based. A flight task is chosen, performance criteria established and the test pilot attempts to perform the task. The test pilot is required to assess his performance in completing the task and assess subjective level of compensation in performing the task. The results lead to the rating levels shown in Table 9.1.

Table 9.1   Performance Ratings

Level	Description	Deficiencies	Performance Level and Pilot Compensation
Level 1	Satisfactory	Improvement not required	Desired performance with minimal pilot compensation
Level 2	Acceptable	Improvements warranted, but not required	Desired performance with high pilot compensation required or Adequate performance with moderate pilot compensation
Level 3	Unacceptable	Improvement required	Performance not adequate with moderate pilot compensation
Level 4	Unacceptable	Improvement required	Performance not adequate with high pilot compensation or Pilot unable to complete task

It is implied that pilot compensation will be high if the desired performance is not attained.

Cooper-Harper Logic Tree

The standard reporting format is the Cooper-Harper (*146*) handling qualities rating (HQR). The HQR scale uses a decision tree (Figure 9.1) to allow the pilot to walk-through a series of dichotomous alternatives, by answering questions, such as "Is it [the aircraft] controllable?"; "Is adequate performance attainable with a tolerable workload?"; And "Is it satisfactory without improvement?" Following these dichotomies, the pilot then makes a choice of three sub-alternatives based on his subjective assessment of the degree of compensation required.

Cooper and Harper emphasized that these ratings be taken in the context of a specific flight task flown by a typical operational pilot. The evaluator must use consistent and clearly stated performance standards.

There are a number of advantages to the Cooper-Harper approach. First, the flow chart involved produces consistent results, particularly with trained evaluators. This is evident in the area of aircraft handling qualities ratings. The second advantage is the direct determination of acceptable results versus problems that require fixes. Other rating scales, such as workload or comparison scales do not provide these direct levels of acceptability.

Further treatment of handling qualities ratings can be found in references (*146*) and (*147*).

Display Readability Rating

There are two aspects of flight displays that must be considered: can the pilot determine the value of a specific parameter, such as airspeed? And can the display be used to control that parameter? These two questions must be answered in the context of a specific task scenario.

Because of the wide-spread acceptance of the HQR scale in the flight test community, two flow charts were constructed to rate the readability and the controllability of displays.(*23*) The display readability chart is shown in Figure 9.2.

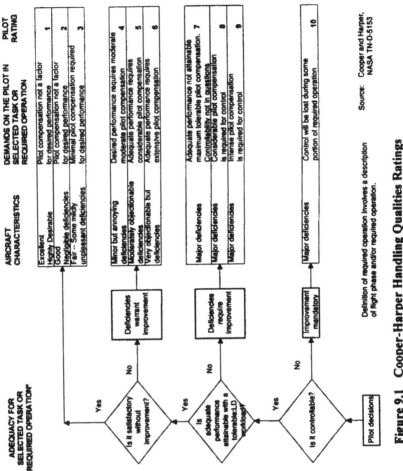

Figure 9.1 Cooper-Harper Handling Qualities Ratings
Source: *Cooper and Harper (146)*

Like the HQR, the display evaluation ratings are flight task dependent. The display readability rating (DRR) requires the evaluator to rate his ability to determine parameter values with desired/adequate accuracy in the context of the flight task being performed. The DRR was validated in a study at Naval Postgraduate School.(*148*) It was also used in the civil certification of a HUD in a Beech King Air.(*9*)

The readability rating can also be applied to the ease of overall maintenance of situation awareness (SA). However, such ratings are only a subjective assessment of SA.

Display Flyability Rating

The display flyability ratings follow the original HQR decision tree closely. The difference between the display flyability rating and a HQR is the requirement that the evaluation pilot consider aircraft control using the display for information. The flow chart is shown in Figure 9.3.

The DFR is essentially a HQR of the airplane handling qualities in series with the display control laws. The HQR is an airframe and flight control rating. For the purposes of display evaluation, the HQR is obtained with the pilot using direct view and traditional steam gauges for flight data. The MTEs used to obtain HQRs are generally tasks based on visual flight (such as ground reference tasks).

DFRs are obtained using the display under test to provide flight data in the same task. HQRs emphasize the airframe response while DFRs rate the combination of airframe and sensor/display system. DFRs can also be used when using a display to control a system other than an aircraft, such as using a display to aim a weapon system. Such a control task falls under the DFR schema.

Like the HQR and the DRR, display flyability ratings are task dependent and the same careful attention must be paid to ensuring that the tasks are appropriate and that suitable performance criteria are established. The DFR requires the evaluator to rate his ability to achieve desired/adequate performance goals and the amount of compensation required to correct for display deficiencies in the context of the flight task being performed.

Display Flight Data Cards

Recommended flight data cards to record the display ratings are shown as Figures 9.4 and 9.5. Figure 9.4 shows the data card for general flight tasks, while Figure 9.5 shows the data card for ILS approaches (adapted from the data card used by Anderson *et al.*(9)

These data cards are designed to minimize the training required for use by operational pilots. The pilots are asked the questions appropriate for the decision branches in the ratings (Figures 9.2 and 9.3). They are also asked to state the degree of reading accuracy and of the controllability. (This latter data will be used in the performance assessment evaluation.)

Choice of Pilots

A significant issue is "What pilots should be used to perform the display evaluations: test pilots or operational pilots?"

Arguments favoring operational pilots include having pilots with recent mission experience. It is also possible to obtain a range of experience levels from recent pilot training graduates to experienced pilots.

One problem with using operational pilots is that each pilot is often overtrained on a particular display and may be predisposed to that display — F-16 pilots prefer F-16 symbology, F-18 pilots prefer F-18 symbology, etc. Ideally, one should use operational pilots with no symbology background. Unfortunately, this is not possible. To avoid this problem, the experimenter must ensure that no particular symbology is over-represented and that the subjective data is used with care.

Another problem is the need to train operational pilots, both in how to fly with non-standard displays or techniques and in how to use rating scales. It is imperative that adequate familiarization and instructions be provided. This is most apparent with scales similar to the HQR. This training can amount to two or three practice sorties per pilot compared with one for a trained evaluator. This problem area can not be overstated and is one of the most severe restrictions on using line pilots.

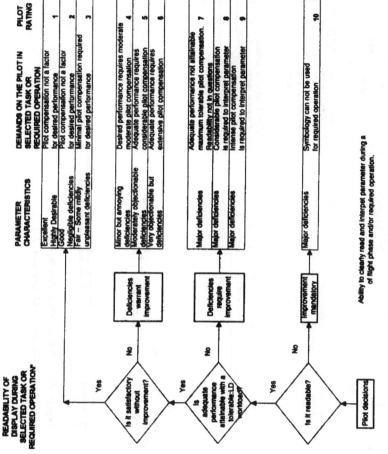

Figure 9.2 Display Readability Rating
Source: Haworth and Newman (23)

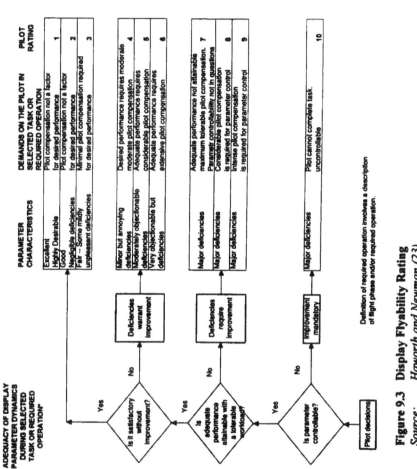

Figure 9.3 Display Flyability Rating
Source: Haworth and Newman (23)

MANEUVER DATA CARD				
PILOT NAME:			SORTIE: _____ DATE:_____	
			TIME: _____	
MANEUVER			CONFIGURATION	

READING DISPLAY	READ DISPLAY TO WHAT ACCURACY?	IS THIS THE DESIRED ACCURACY?	EFFORT REQUIRED FOR THIS ACCURACY	REMARKS (include deficiencies, degree of pilot compensation, and possible improvements.)
PITCH		☐ Desired ☐ Adequate ☐ Not Adequate ☐ None	☐ None ☐ Minimal ☐ Moderate ☐ Considerable ☐ Excessive	
ROLL		☐ Desired ☐ Adequate ☐ Not Adequate ☐ None	☐ None ☐ Minimal ☐ Moderate ☐ Considerable ☐ Excessive	
YAW/ HEADING		☐ Desired ☐ Adequate ☐ Not Adequate ☐ None	☐ None ☐ Minimal ☐ Moderate ☐ Considerable ☐ Excessive	
AIRSPEED/ AOA		☐ Desired ☐ Adequate ☐ Not Adequate ☐ None	☐ None ☐ Minimal ☐ Moderate ☐ Considerable ☐ Excessive	
OVERALL		☐ Desired ☐ Adequate ☐ Not Adequate ☐ None	☐ None ☐ Minimal ☐ Moderate ☐ Considerable ☐ Excessive	

CONTROL	CONTROL PRECISION ACHIEVED	IS THIS THE DESIRED TOLERANCE?	EFFORT REQUIRED FOR THIS CONTROL	REMARKS (include deficiencies, degree of pilot compensation, and possible improvements.)
PITCH		☐ Desired ☐ Adequate ☐ Not Adequate ☐ None	☐ None ☐ Minimal ☐ Moderate ☐ Considerable ☐ Excessive	
ROLL		☐ Desired ☐ Adequate ☐ Not Adequate ☐ None	☐ None ☐ Minimal ☐ Moderate ☐ Considerable ☐ Excessive	
YAW/ HEADING		☐ Desired ☐ Adequate ☐ Not Adequate ☐ None	☐ None ☐ Minimal ☐ Moderate ☐ Considerable ☐ Excessive	
AIRSPEED/ AOA		☐ Desired ☐ Adequate ☐ Not Adequate ☐ None	☐ None ☐ Minimal ☐ Moderate ☐ Considerable ☐ Excessive	
OVERALL		☐ Desired ☐ Adequate ☐ Not Adequate ☐ None	☐ None ☐ Minimal ☐ Moderate ☐ Considerable ☐ Excessive	

Figure 9.4 Maneuver Data Card

APPROACH PILOT DATA CARD				
PILOT NAME:		SORTIE: _____ DATE:_____		
		TIME: _____		
AIRPORT/APPROACH		FLAPS: _____ Flight Director (_) Yes (_) No		
		VREF: _____		

READING DISPLAY	READ DISPLAY TO WHAT ACCURACY?	IS THIS THE DESIRED ACCURACY?	EFFORT REQUIRED FOR THIS ACCURACY	REMARKS (include deficiencies, degree of pilot compensation, and possible improvements.)
PITCH		(_) Desired (_) Adequate (_) Not Adequate (_) None	(_) None (_) Minimal (_) Moderate (_) Considerable (_) Excessive	
ROLL		(_) Desired (_) Adequate (_) Not Adequate (_) None	(_) None (_) Minimal (_) Moderate (_) Considerable (_) Excessive	
YAW/ HEADING		(_) Desired (_) Adequate (_) Not Adequate (_) None	(_) None (_) Minimal (_) Moderate (_) Considerable (_) Excessive	
AIRSPEED/ AOA		(_) Desired (_) Adequate (_) Not Adequate (_) None	(_) None (_) Minimal (_) Moderate (_) Considerable (_) Excessive	
DISTANCE TO GO		(_) Desired (_) Adequate (_) Not Adequate (_) None	(_) None (_) Minimal (_) Moderate (_) Excessive	
OVERALL		(_) Desired (_) Adequate (_) Not Adequate (_) None	(_) None (_) Minimal (_) Moderate (_) Considerable (_) Excessive	

CONTROL	CONTROL PRECISION ACHIEVED	IS THIS THE DESIRED TOLERANCE?	EFFORT REQUIRED FOR THIS CONTROL	REMARKS (include deficiencies, degree of pilot compensation, and possible improvements.)
PITCH		(_) Desired (_) Adequate (_) Not Adequate (_) None	(_) None (_) Minimal (_) Moderate (_) Considerable (_) Excessive	
ROLL		(_) Desired (_) Adequate (_) Not Adequate (_) None	(_) None (_) Minimal (_) Moderate (_) Considerable (_) Excessive	
YAW/ HEADING		(_) Desired (_) Adequate (_) Not Adequate (_) None	(_) None (_) Minimal (_) Moderate (_) Considerable (_) Excessive	
AIRSPEED/ AOA		(_) Desired (_) Adequate (_) Not Adequate (_) None	(_) None (_) Minimal (_) Moderate (_) Considerable (_) Excessive	
OVERALL		(_) Desired (_) Adequate (_) Not Adequate (_) None	(_) None (_) Minimal (_) Moderate (_) Considerable (_) Excessive	

Maximum deviation between: Lateral:_____
between FAF and minimums: Vertical:_____
Airspeed:_____

Within safe landing window at minimums? (_) Yes (_) No (_) N/A

Landing or Missed Approach? (_) Straight-in landing (_) Circle to runway ___
(_) Planned missed approach
(_) Missed approach, pilot decision
(_) Missed approach: reason:_____

Figure 9.5   Approach Data Card

Arguments favoring test pilots include having trained evaluators. Properly test pilots are used to rating airplane handling and should be familiar with the rating scales, such as the Cooper-Harper type of walk-through ratings. Test pilots are also skilled at communicating with engineers and can provide insight into display or control law problems.

The need to conduct practice sorties for untrained evaluators can quickly use up the available sorties in a program. For example, if 24 sorties are available, using two test pilots will allow for twenty-two data sorties. If six operational pilots are used instead, twelve to eighteen practice sorties may be required allowing only six to twelve data flights. This should not be an issue for simulator trials — only for flight test.

Initial Testing

Test pilots should be used to perform the evaluations for the initial display verifications, calibrations, and dynamics testing.

Test pilots with experience in evaluating aircraft dynamics and training in methodical evaluations should be used for the initial series of testing: Verification that the display meets the design specifications and the initial display dynamics evaluations. The use of test pilots for the initial dynamics evaluations, open-loop testing and aggressive MTEs, should ensure that any dynamics issues are addressed immediately.

Mission Task Element Display Evaluations

A combination of test pilots and operational pilots should perform the evaluations for both precise and aggressive MTE testing. The test pilots would serve to bridge the gap between HQ testing and the display evaluations, while the operational pilots bring their mission experience and provide a spectrum of experience. It is important that the operational pilot experience include diversity of experience.

For most display *flight* evaluations, two test pilots and two to four operational pilots should be a reasonable compromise between a sufficient statistical sample and the typical cost and time constraints associated with a display evaluation program.

Mission Testing

Four to six operational pilots should perform the evaluations for mission related testing and situation awareness testing. These pilots should be current in the mission or a similar mission. Normally, test pilots would not be appropriate for these evaluations,* although they should serve as safety pilots and have inflight responsibility for ensuring that test conditions are met.†

The limitations that budget constraints place on test sorties in the aircraft lead to the need for extensive simulation trials. As many pilots as possible should be used for mission testing to ensure a valid statistical sample.

Testing in Reduced Visibility

Display testing requires flight in a variety of environmental conditions. Many see-through display issues arise because there is additional information (the real world) visible through and around the symbolic information. In solid IMC, this has no effect because the pilot could not see the real world. In clear weather, there is minimal effect because the real world cues overwhelm the symbolic cues. The problem arises when the external cues are faintly or partially visible and may be misleading. Garman and Trang (*135*) reported a significant deterioration in suitability when a head-mounted display was in poor visibility compared with flight in good visual conditions.

To date, most see-through display testing has been conducted in either extreme of visibility, solid IMC or severe VMC. The in-between conditions may be the more severe, from a display point-of-view because of misleading cues. They are also the most difficult to obtain.

---

* By excluding test pilots, we mean to exclude company and authority project pilots who have been working on the systems. We don't mean that because a pilot has been a test pilot at some point in his career, he should not participate in mission testing. We do want these evaluation pilots to be currently flying operationally as their primary duty.

† For flight tests in two-pilot aircraft, the second pilot will normally be a project test pilot.

Simulating Instrument Conditions

The traditional means of restricting a pilot's visual cues, for training or evaluation, has been the instrument hood (the "bag"). Hoods have been used since the earliest days of instrument flight. This is an easy and effective means of producing solid instrument conditions. Problems with instrument hoods are restrictions to safety pilot's view and difficulty in quickly restoring visual conditions. Tandem seat aircraft usually require the hooded pilot to sit in the rear seat; otherwise the safety pilot's view would be blocked.

Complementary-color filters have been used to block external visual cues with minimal restrictions to the safety pilot's view. The most common form is blue/amber. The windshield and canopy or side windows are covered with an amber sheet of plastic. The view through the amber film is colored, but does not prevent the safety pilot from seeing outside. The complementary color method is effective and allows the external vision to be rapidly regained to simulate breakout on an instrument approach. Like the hood, It is an all-or-nothing method.*

Simulating Degraded Visual Environment

Hoh (*149*) evaluated "foggles" that are goggles with ground surfaces to block high-spatial-frequency images allowing low-spatial-frequency information to pass. This technique is promising for daytime degraded visual conditions, but has not yet been tried under night conditions.

Many of these conditions can be produced in some form in simulators. Other than the solid instrument conditions, simulator visual scenes only approximate the real world.

---

* A note of caution is in order. In one instance, red/green filters were used in place of blue/amber. The windshield and side windows were covered with red plastic film and the evaluation pilot wore a green visor. As it happened, the safety pilot on one occasion was flying under a medical waiver for red-green color-blindness. He commented that he could not tell that the red plastic was installed as it appeared perfectly transparent. During the flight, the sun went behind a cloud and the ambient brightness decreased considerably. This caused the red plastic to suddenly become opaque. It is not conducive to a test for the safety pilot to suddenly shout "I can't see anything!"

Flight in Actual Conditions

Finally, flight in actual instrument conditions is an option. However, this presumes a fully qualified instrument suite and supporting systems for the safety pilot. Such qualification is usually not available for new aircraft designs.

An FAA/USAF technology demonstration program for synthetic vision flew actual low visibility approaches in a Gulfstream to determine the acceptability of using FLIR and MMWR to create runway images.(*150-151*) The approaches were flown in actual low-visibility conditions down to 50 ft DH. As the image was recorded, the airplane carried air-sampling sensors to characterize the inflight liquid water content, water droplet size and distribution, air temperature, etc. The airport environment was surveyed and calibrated to establish radar cross sections for ground features. This allowed correlation of the actual atmospheric conditions with the recorded image quality and pilot responses. It also supported correlation of empirical data with theoretical performance prediction. To our knowledge, this is the only structured flight test program studying degraded visual conditions in actual weather. All of the data collected during these tests was organized and recorded in a database in digital and video formats and is available today.*

Data Considerations

Internal Aircraft State

There is one significant advantage to modern displays and systems. It is no longer necessary to generate many square yards of oscillograph tracings to be transcribed after the flight. It is much easier to obtain time histories of internal parameters by means of digital recording. In most new aircraft, the internal aircraft state is available on one or more databuses. "All that is needed," is for the flight test engineer to develop a data collection methods. This is often done with flight test software loads which incorporate data collection or recording schemes internally. The alternative is to use a bus analyzer.

---

* Malcolm Burgess (Research Triangle Institute), personal communication, 2000.

External Aircraft State

By external aircraft state, we mean the air mass data (airspeed, barometric altitude, angle-of-attack, etc.), the orientation (pitch, roll, yaw, or heading), the position in three dimensions, and velocities. Normally these can be obtained by recording internal variables, such as air-data-computer output, or gyro or AHRS output.

Tracking

GPS, in particular, can provide accurate position in three dimensions, inertial velocities in three dimensions, and accurate time. In fact, GPS time should be considered as the most suitable means for time stamping all data from different sources (i. e. ground tracking and on-board data). Ground tracking data, with embedded GPS time and position, can be combined with on-board GPS data to provide an ex-post-facto differential GPS which will likely be as good or better than any tracking radar data.

Helicopter positions, particular as used in the ground-referenced MTEs, require more accurate position data than is normally available on board the aircraft. In the past theodolite tracking was used. Laser (or LIDAR) tracking may be more suitable although this will usually require a laser retro-reflector mounted on the exterior of the aircraft to define a specific location. Differential GPS (DGPS) is another alternative to optical or radar tracking.

Range survey data: Frequently, it is necessary to track aircraft position relative to ground targets, markings, or features. Examples of such tests include approach and landing evaluations, air-to-ground targeting, ground proximity warning systems (GPWS), or terrain following/terrain avoidance (TF/TA) systems. For such tests, the test range or airport environment must be accurately surveyed and the location of obstacles, runways, ground markings, and significant terrain features noted.

During the testing, position relative to these features must be recorded. Traditionally, theodolite data or tracking radar has been used. GPS data is probably a better approach. Even if true differential GPS is not used, accurate relative position can be obtained. Locating a GPS receiver with recording capability at a fixed ground location will allow the flight test engineer to combine GPS data from multiple sources and develop pseudo-differential GPS positions. For fixed ground locations, the equipment can be a simple as a handheld receiver and a laptop computer.

A sample ground-range test procedure follows in the next subsection (page 110).

Cooperative Aircraft Tracking: Depending on the system under test, it may be necessary to include cooperative aircraft in the test scenario. Such tests could include evaluations of air-to-air radar, threat detection equipment, traffic alert/collision avoidance system (TCAS), or stationkeeping equipment (SKE). For such tests, relative locations of the test aircraft and other cooperative aircraft must be determined and recorded as a function of time.

In some cases, such as air-to-air visual tracking, on-board photography or video recording is suitable. In others, tracking will be more involved. Radar tracking has been used in the past. However, individual GPS recording on each aircraft provides sufficient data as well as timing information. GPS can also incorporate aircraft track and velocities. If aircraft attitude information is recorded simultaneously, target aspect information can be provided at the same time.

A sample cooperative aircraft test procedure follows in the next subsection (page 112).

Data Sampling

Sampling rates should be frequent enough to ensure that the bandwidth of interest is recorded. The basic rule-of-thumb is to sample at 2-4 times the highest frequency. As a matter of practice, however, sampling will be driven by the data bus rates. Digital data will be sampled from the data bus at the bus rate (or its submultiple).

Sampling at too great a rate can lead to storage problems and excessive post-test processing requirements. Sampling at too small a rate can lead to loss of high-frequency data. The flight test engineer must ensure that neither occurs. Table 9.2 shows typical data sampling rates. Specific aircraft and system issues may require greater sampling rates.

MIL-STD-1553 bus: If the data is sampled from a 1553 bus, flight test software for the bus controller will have to be used to transmit the data to the data recording terminal. MIL-STD-1553 (*152*) stipulates that a bus terminal will only respond to data specifically addressed to it.

Table 9.2 Normal Data Sampling Rates

Type of Data	Normal Sampling Rate
Air Mass (airspeed, altitude, etc.)	4 Hz
Gyro Attitudes and Rates	20-40 Hz
Heading (up and away flight)	10 Hz
Heading (ground reference)	20-40 Hz
Enroute Position	1 Hz
Ground Reference Position	10 Hz
Angle of Attack/Sideslip Vanes	1-4 Hz
Radar Altitude	4 Hz
Sight Picture	30 Hz

ARINC-429 bus: If the data is sampled from a 429 bus, no flight test software for the bus controller will be required, since ARINC-429 (*153*) is a broadcast bus protocol.

Analog-to-digital recording: If the data is obtained from analog sources, such as pressure transducers, force gauges, etc. analog/digital (A/D) conversion will be necessary. In such cases, the flight test engineer will have to ensure that adequate sampling is performed and the data recorded.

Display Presentation

There is often a need for recording the display itself. Most have some form of video output which can be recorded. This is particularly important for see through displays, such as HUDs or HMDs. Fortunately, most HUDs incorporate provisions for data-plus-external scene recording.

Sample Flight Tests

We have two flight evaluations to outline procedures and precautions for conducting tests involving ground and airborne threats. These tests are part of the verification of the Ground Proximity Warning System (GPWS) and the Traffic Advisory and Collision Warning System (TCAS).

Terrain Closure Test (GPWS)

The verification of GPWS functionality is typical of similar systems, ground collision avoidance systems (GCAS), terrain awareness warning systems (TAWS), or enhanced ground proximity warning system (EGPWS). The approach is also used in tests of air-to-ground radar systems.

GPWS is a warning system with several operating modes providing warning based on absolute altitude, rate of descent, closure rate to terrain, and aircraft configuration. The warning modes are

    i *excessive sink rate vs. absolute altitude*
    ii *excessive closure vs. absolute altitude*
    iii *descent after takeoff*
    iv *too low for configuration*
    v *descent below glideslope.*

Details of the warning envelopes are described in an advisory circular.(*154*) In this section, we shall discuss guidelines for performing the first two tests.

a **Objectives**: The objective of this evaluation is to verify that the warning response matches the specifications.

b **Requirements**: The warning electronics and sensor(s) must be fully functional. The actual cockpit warning need not be functional. Radar tracking (or differential GPS) is required.

    i excessive sink rate vs. absolute altitude: This test requires flat terrain.

    ii *excessive closure vs. absolute altitude*: A surveyed test range with suitably rising terrain must be available. Normally, a ridge line should provide a suitable closure rate with the aircraft in level flight (at an altitude that will clear the ridge). The objective is to fly toward the rising terrain at an altitude designed to trigger the warning while providing adequate terrain clearance to minimize risk to the test aircraft and crew.

c **Action**

    i *excessive sink rate vs. absolute altitude*: The pilot will descend to a predetermined minimum altitude at varying rates-of-descent. Normally, the procedure is to fly at a low rate-of-descent until receiving the warning or a minimum safe altitude. The test should be repeated at increasing descent rates until sufficient data is obtained.

ii *excessive closure vs. absolute altitude*: The pilot will fly toward the ridge at varying altitudes. Normally, the procedure is to fly at an altitude that will not trigger the warning, and gradually lower the altitude until the warning occurs.

d **Data**: Aircraft state data shall be recorded as time histories. Radar tracking or GPS time histories should be recorded. Locating a GPS receiver on the test range will allow *ex post facto* differential GPS data for position.

e **Performance criteria**: The warning signal should match the specifications.

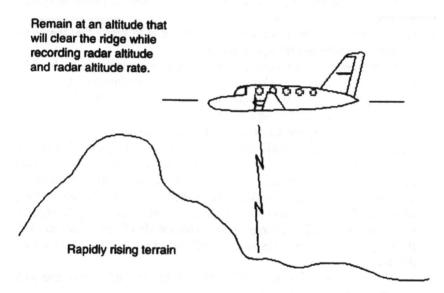

Remain at an altitude that will clear the ridge while recording radar altitude and radar altitude rate.

Rapidly rising terrain

Figure 9.6   Terrain Closure Test

f **Safety considerations**: This is a hazardous test and requires careful attention to minimum safe altitudes and benign weather conditions (good visibility and light winds). The pilot must maintain visual contact with the terrain and ridge. Tracking data should be monitored and the pilot warned if test safety criteria are penetrated.

Threat Detection for Approaching Airplane (TCAS)

The verification of TCAS functionality is typical of similar threat detection systems. The approach is also used in tests of air-to-air radar systems.

TCAS is a warning system which monitors radar transponder replies from nearby aircraft and, depending on proximity and rate of closure, presents traffic advisories (TA's) or resolution advisories (RA's). RA's command collision avoidance maneuvers. In the present implantation of TCAS (TCAS-II), only vertical maneuvers are used.

Details of the warning envelopes are described in an advisory circular.(*155*) In this section, we shall discuss guidelines for performing the a typical test.

a **Objectives**: The objective of this evaluation is to verify that the warning response matches the specifications.

b **Requirements**: The warning electronics and sensor(s) must be fully functional. The actual cockpit warning need not be functional.

   Flight along a linear feature, such as a road or section line can assist the pilots in maintaining a correct track.

   Radar tracking (or differential GPS) is required.

c **Action**: The pilots should fly predetermined flight paths that should bracket the warning envelope boundaries. As an example, consider the head-on case. The pilots should fly approaching tracks at altitudes separated by 1000 feet. In this case, traffic advisories, but no warning (RA) should occur. This should be repeated at successively lower altitude spacing to until warnings occur. The test should be continued with the target aircraft below the test aircraft until no RA's are no longer indicated.

   The test should be repeated with varying rates of closure and with the aircraft on crossing tracks.

d **Data**: Aircraft state data shall be recorded as time histories. Radar tracking or GPS time histories should be recorded. Locating a GPS receiver on the test range will allow *ex post facto* differential GPS data for position.

e **Performance criteria**: The warning signal should match the specifications.

f **Safety considerations**: This is a hazardous test and requires careful attention to avoiding collisions and good flight visibilities. The pilots should maintain communication with each other and a minimum separation boundary established that may not be penetrated without visual contact. Flight along a linear terrain feature, such as a road or section line can help maintain inflight separation.

Small differences in altitude (of the order of fifty feet) can help assure separation, if such a vertical amount will not compromise the purpose of the test,

Tracking data should be monitored by a safety observer and the pilots warned if test safety criteria are penetrated.

10 Initial and Collateral Tests

Laboratory Tests

There are a number of laboratory tests which are required prior to flight release. These include ensuring that the equipment can withstand the flight environment and will not interfere with (or be interfered by) other equipment on the airplane. In addition, the equipment performance must be established in a controlled laboratory environment. It is not our intent to discuss these tests since they are common to all electronic equipment. Spitzer's book (*156*) discusses these tests. Further details for environmental testing is found in references (*80* through *85*).

We do wish to included some optical measurements which are unique to head-up displays (HUDs) or helmet-mounted displays (HMDs). These have been included for two reasons. First, they are relatively unknown outside of the HUD manufacturers. Second, they give some insight into HUD/HMD specifications and what they mean to operations and to test and evaluation.

HUD Optical Measurements

HUD optical measurements are generally accomplished in an optical laboratory because of the need for precise positioning of the theodolite. While measurement in the airplane is possible, it is not normally accomplished.

These tests include the following.
   i   *field-of-view (FOV)*
   ii *binocular parallax*
   iii *HUD eyebox verification*
   iv *symbol positioning accuracy*
   ·v *real-world distortion*

Details of these and other tests can be found in SAE ARP-5287.(*76*)

a **Objectives**: The objective of this evaluation is to verify that the HUD optical properties match the specifications.

114

b **Requirements**: HUD fields of view are generally measured using a theodolite. A theodolite is a precision optical instrument capable of measuring azimuth and elevation angles. The theodolite precision should be .05 mrad* or better. It generally consists of a medium power telescope mounted to a mechanical assembly allowing vertical and horizontal rotations. The entrance aperture of some theodolites is about 50 mm. Since the entrance aperture of an eye is about 6 mm, for some measurements the entrance aperture of the theodolite should be stopped down using an aperture mask to about 6 mm.

The theodolite is positioned such that its pivot axis is located at the eyebox position where the measurement is to be taken. Thus, when measuring the HUD instantaneous FOV from the center of the HUD Eyebox, the pivot point of the theodolite must be positioned at the center of the eyebox. This requires the theodolite to be mounted to a three-axis translation stage assembly with adequate motion range to cover the eyebox volume.

The basic measurement technique assumes a suitable test pattern fills the HUD total field of view. An example of a suitable test pattern is a grid made up of 7 horizontal lines intersecting 7 vertical lines. Using an odd number of lines provides a center reference. If not located at the center of the HUD FOV, a specific boresight symbol should be displayed.

c **Action**

  i *field-of-view(FOV)*: Locate the theodolite at the desired HUD eyebox location using the translation stage assembly.

Rotate the theodolite in both azimuth and elevation in order to view one end of one of the horizontal lines. Record both the horizontal and vertical values. Slew the theodolite horizontally to read the other end of the same horizontal line, and record these horizontal and vertical readings. The difference in horizontal reading represents the angle subtended by the horizontal line. Repeat the measurement for each horizontal line. Repeat the measurements for each vertical line.

Repeat these steps for each eyebox position of interest necessary to characterize the HUD FOV.

---

* nominally 10 arc sec.

Once the instantaneous monocular FOV data is collected from the desired eyebox positions, it can be reduced to determine the various fields of view (total FOV, instantaneous FOV, and binocular overlapping FOV). It is often useful to generate FOV plots for each eyebox location evaluated.

The total FOV is determined by computing the maximum horizontal and vertical display limits visible from any position tested within the HUD eyebox. Thus, if the instantaneous monocular FOV from only one head position extends from -16 degrees to +16 degrees horizontally, then the total horizontal FOV is a minimum of 32 degrees.

The HUD instantaneous FOV for a given head position within the HUD eyebox is determined by computing the maximum horizontal and vertical display limits from two eyebox locations horizontally spaced apart by 2.5 in. Thus, the instantaneous FOV from 1.25 in above the design eye is determined from the instantaneous monocular FOVs from eye locations (up 1.25 in, left 1.25 in), and (up 1.25 in, right 1.25 in).

The binocular overlapping FOV for a given head position within the HUD eyebox is computed by determining the common horizontal and vertical field angles from two eyebox locations horizontally spaced apart by 2.5 in. Thus, the overlapping binocular FOV for a head positioned (up 0.5 in, right 1.0 in) is determined by computing the common field angles from two eyes positioned at (up 0.5 in, left 0.25 in) and (up 0.5 in, right 2.25 in).

By analyzing the monocular FOV data from several carefully selected locations within the eyebox, the HUD FOV characteristics can be determined.

ii *binocular parallax*: Determining HUD binocular parallax requires two angular measurements of the same display element to be made for each field angle/head position of interest. The first measurement is made from the left eye position, and the second measurement is made from the right eye position. The binocular parallax value for a specific field angle and head position is the difference between the two measurements. Horizontal and vertical errors are characterized independently.

Parallax measurements are generally made using a theodolite using methods similar to those described for field of view measurements. A minimum of 9 field points from each of five head positions within the HUD eyebox is suggested to fully characterize the HUD binocular parallax. The origin of the coordinate reference frame discussed below is the center of the HUD eyebox.

The following summarizes one approach to determining binocular parallax:

Determine which field angles and eyebox head positions are to be tested. Determine the eye box positions for the left eye and right eye for each eyebox position to be measured. For example, the eyebox positions for measuring the parallax from one inch to the left would be (up 0, left 2.25 in) and (up 0, right 0.25 in).

Locate the theodolite pivot point at the desired left eye position using the adjustable stage assembly.

Point the theodolite at each field point of interest. Record the horizontal and vertical theodolite scale readings for each field position.

Translate the theodolite to the desired right eye position using the adjustable stage assembly and repeat the measurement.

The horizontal and vertical parallax values for each field point by are the differences between the left eye values from the right eye values.

iii *HUD eyebox verification*: The physical dimensions of the HUD eyebox are determined in a manner similar to determining the HUD instantaneous monocular field of view. The HUD specifications will describe a minimum monocular FOV which must be visible for a location to be considered within the eyebox. To show compliance with this requirement, a theodolite is positioned at each edge of the three dimensional eyebox specified for the HUD. The operator measures the instantaneous monocular FOV using the methods described above. The vertical and horizontal limits of the instantaneous monocular FOV are determined from the eyebox limits and compared with the specified minimum requirement.

iv *symbol positioning accuracy*: HUD accuracy can be determined by measuring the symbol locations as generated by the HUD processor and displayed on the display unit, and by analytically accounting for the installation accuracy errors. Alternatively, symbol positions could be determined by measuring the display unit (with a suitable test pattern), and analytically including both the processor and installation error components. Either method is acceptable. Assumptions about processor error components must be justified.

The symbol positioning errors are determined using a theodolite located at the eyebox center. The theodolite scales provide an absolute angular reference frame to which the location of HUD symbols can be compared. Thus, the angular locations of display elements with respect to a reference symbol can be measured and compared with the desired location. The total accuracy error is then computed by numerically summing the display unit errors with the installation and processor errors (if applicable). The theodolite should be equipped with a 6 mm diameter aperture mask when performing these tests.

v *real-world distortion*: Light refraction through non-flat HUD combiners can introduce real world distortions and parallax errors. Combiner induced real world distortion is the change in angular position or size of real world objects as viewed through the combiner compared with the real world viewed without the combiner. Combiner Induced real world parallax errors are binocular disparity errors of the real world view introduced when viewing the real world through a combiner. There are no combiner induced distortion or parallax errors when the HUD combiner surfaces are flat and parallel. The worst case combiner induced real world distortions for curved combiners generally occur near the edge of the eyebox. Real world distortions and parallax errors can be measured using a theodolite.

It is often difficult to measure objects at or near optical infinity in a laboratory environment. One method to simulate optical infinity is to autocollimate a theodolite from the surface of an optically flat mirror. In this method, the collimated illuminated cross hairs of the theodolite are reflected from a flat reference mirror and are returned down the optical axis of the theodolite to form an inverted image of the cross hairs. If the theodolite is positioned to be exactly perpendicular to the reference mirror, the cross hair image will exactly overlay the real cross hair. Under these conditions, the theodolite is said to be autocollimated.

The general procedure for measuring real world distortions is to introduce the combiner into the autocollimated beam from the theodolite. The theodolite is re-autocollimated (i.e., the theodolite angles are slightly adjusted) and the angular deviation due to the combiner is determined. Real world distortions and line of sight deviations are generally largest where the combiner curvature is greatest. Thus, this measurement should be made from the vertical and horizontal limits of the eyebox, and from the center of the eyebox. The worst case look angle for distortion generally occurs at the physical edges of the combiner glass. Thus, these measurements should be made from the edges of the eyebox looking through the limits of the combiner glass assembly, and from the center of the eyebox looking at the center of the display. The following is the general process for making these measurements.

Determine the eyebox eye positions to be evaluated. Determine the field angles necessary to view through the edges of the combiner element from each eyebox position.

Locate the theodolite pivot point at the desired eye position relative to the HUD combiner using the adjustable stage assembly.

Point the theodolite at the desired field angle. Position and adjust the reference mirror such that the theodolite is autocollimated on the mirror. Record the horizontal and vertical theodolite scale readings.

Locate the combiner in the correct relative position to the theodolite between the theodolite and the reference mirror. Re-autocollimate the theodolite on the reference mirror. (Do not move the reference mirror relative to the theodolite base.) Record the horizontal and vertical theodolite scale readings.

Calculate the change in autocollimation readings. The change is the line of sight errors through the combiner.

Repeat the measurements for each eye position/field angle necessary to characterize the combiner. Combiner distortions are the change in line of sight errors with respect to the line of sight error at the center of the eyebox looking at the center of the display.

Combiner induced real world parallax errors are determined by computing the difference in the line-of-sight errors from two eyebox locations displaced horizontally by 2.5 in, the interpupillary distance. The measurement method is the same as for determining distortion, except that eyebox locations are selected for head positions, as opposed to eye positions. Note that the reference mirror must be wide enough to allow autocollimation from each eye position without repositioning the mirror.

d **Data**: The data consists of angular measurements (horizontal and vertical) obtained from the theodolite and the position of the theodolite (x, y, and z coordinates) relative to the DERP.*

e **Performance criteria**: The HUD optical characteristics should comply with the specifications. Normally, disparity tolerances are expressed as two-sigma values (i. e., Ninety-five percent of the measurements should be within specified tolerances).

f **Safety considerations**: There are no particular hazards associated with this laboratory test.

HUD Photometric Measurements

HUD of photometric measurements measure the brightness of the symbols or images displayed on a HUD or HMD. These are generally accomplished in an optical laboratory. They include the following:

 i *line width*
 ii *luminance*
 iii *combiner transmissivity*

Details of these and other tests can be found in SAE ARP-5287.(76)

a **Objectives**: The objective of this evaluation is to verify that the HUD optical properties match the specifications.

b **Requirements**: HUD photometric measurements are generally made using a spot photometer capable of measuring display luminance over a viewing angle of less than the display line width.

c **Action**:
 i *line width*: The normal definition of line width is based on the distance between the fifty-percent intensity points on either side. The theodolite could be used to make this measurement. However the difficulty in reproducing the fifty-percent points using subjective assessment makes this approach unsatisfactory.

 A slit with a angular width of about ten percent of the HUD line width is slowly scanned across a displayed line (the slit is aligned to

---

* The Design Eye Reference Point (DERP) is the point relative to the HUD used in the optical design. The Design Eye Point (DEP) is the point in the cockpit where the pilot's eyes are presumed to be and from which the pilot must have adequate reach and view.

be parallel with the display line). A sensitive detector is positioned behind the slit. The output of the detector is measured as a function of angular slit position. From the data, the fifty percent intensity points can be determined. From this data, the scan angle between the fifty percent intensity points is determined. (A variation of this method is to scan the HUD displayed line across a fixed detector.)

Normally, only data from the DERP is taken (i. e., from the center of the eyebox).

ii *luminance*: HUD photometric measurements are generally made using a spot photometer capable of measuring display luminance over a viewing angle of less than the display line width. The display luminance varies with eyebox position and field angle. To measure the display luminance from a given eyebox position, the photometer entrance aperture must be correctly positioned for each measurement. (In general, the photometer pivot point is not coincident with the measuring axis, so the photometer must be repositioned for every field angle measured.)

iii *combiner transmissivity*: The HUD combiner transmissivity is measured using a photometer located at the center of the display eyebox. This measurement is made against a photopic light source* as follows:

Measure the background luminance.

Position the combiner in the correct spatial relationship with respect to the photometer as the combiner is to the eyebox center when the HUD is installed into an aircraft. Measure the apparent background luminance again.

Combiner transmissivity uniformity is determined by repeating the measurements from at least 5 different eyebox locations.]

d **Data**:
  i *line width*:   A set of line widths at various locations;
  ii *luminance*:   A set of luminance values at various angles and locations within the eyebox;

---

* The transmissivity is usually specified for a photopic spectrum. Photopic refers to the sensitivity of the human eye in bright conditions. The daylight spectrum is approximately the same and is often specified.

    iii *transmissivity*:   Background luminance, L_B
                             Luminance through combiner, L_T
                             Combiner transmissivity, T

$$T = L_T/L_B \tag{10.1}$$

e **Performance criteria**: The HUD optical characteristics should comply with the specifications.

f **Safety considerations**: There are no particular hazards associated with this laboratory test.

Ground Tests

These are preliminary tests, performed on the ground designed to ensure that the cockpit systems meet their specifications and requirements, and that there are no hazards to flight. To minimize program risk, these tests should be conducted in cockpit mockups (egress testing and form and fit testing) or in laboratory tests (symbol list testing) prior to installation in the aircraft. Form and fit testing and symbol list testing should also be performed in simulator cockpits prior to conducting simulator tests. Most of the ground tests should have been evaluated in laboratory mockups to avoid embarrassments when crewmembers first try to enter the cockpit.

HUD Boresight Verification

The alignment of the HUD with the aircraft coordinates requires specific verification in the aircraft. Boresighting tolerances are usually 5 mrad or less. During installation of a HUD into a cockpit, the manufacturer's procedures should ensure that the HUD is properly aligned with the aircraft's longitudinal axis. Following any installation (or when boresight errors are suspected), the boresight accuracy should be verified.

a **Objectives**: The objective of this evaluation is to verify that the HUD installation meets the alignment specifications.

b **Requirements**: Boresighting a HUD requires a fixture to accurately align HUDs are often based on lining up specific rivet heads or using levels on bulkheads. These are not accurate enough for alignment of HUDs. Often a fixed telescope or theodolite is installed on the HUD mount which can be sighted on a distant (at least 60 ft). This allows locating a distant point which is accurately oriented relative to the airplane body axes.

Some HUDs use external boresight alignment fixtures provided for alignment of inertial platforms or AHRS installations.

c **Action**: First, locate an external point using the boresighting telescope installed in place of the HUD. This point should be at least sixty feet from the HUD. A spot on a hangar will is often used.

Reinstall the HUD. Using the boresight symbol (or other test pattern) measure the difference between the HUD boresight symbol and the previously measured spot. An ground technician should move a spot until the pilot indicates that it is covered by the HUD boresight symbol.

d **Data**: The data consists of two distance measurements: The distance between the reference spot and the spot indicated by the HUD boresight symbol (D_S) and The distance from the cockpit DEP to the hangar wall (D_W). Angular error $= \varepsilon = \arctan(D_S/D_W)$.

e **Performance criteria**: The HUD boresight error shall be within specifications.

f **Safety considerations**: There are no particular hazards associated with this ground test.

Head-tracker System Verification

The alignment of the HMD with the aircraft coordinates requires specific verification in the aircraft. Head-pointing tolerances are usually 5 mrad or less. During installation of a head-tracking system (HTS) into a cockpit, the manufacturer's procedures should ensure that the head-tracker is properly aligned and can map the pilot's head motion.

In particular, magnetic head-trackers require extensive mapping of the cockpit magnetic fields. Magnetic HTS's are susceptible to changes caused by new equipment in the cockpit. Following any such installation (or when HTS errors are suspected), the head-tracking accuracy should be verified.

a **Objectives**: The objective of this evaluation is to verify that the head-tracker system (HTS) installation meets the accuracy specifications.

b **Requirements**: The aircraft must be aligned with reference to external visual targets. These must be sufficiently far away to minimize errors based on pilot head position. Assuming a pilot head motion of ±8 inches and a required tolerance of 1 mrad, the marker must be 666 ft from the aircraft. The pilot must be allowed full range of head motion to cover the specified head-motion box.

Celestial objects are quite suitable for above horizontal measurements.

c **Action:** Locate an external point using the boresighting reticle on the HMD. Record the indicated azimuth and elevation. (The HMD display may indicate the azimuth/elevation directly.

Repeat as necessary to cover the full range of external field-of-regard and of the head-motion box. Where tilt correction is included, the pilot should tilt her head left and right for each measurement.

d **Data:** The data consists of a series of elevation and azimuth measures. The pilot's head position within the head-motion box should also be recorded.

e **Performance criteria:** The head-tracker accuracy shall be within specifications.

f **Safety considerations:** There are no particular hazards associated with this ground test.

HUD Field-of-view Verification

While the HUD field-of-view (FOV) was measured in the laboratory, verification of the installed FOV is required.

a **Objectives:** The objective of this evaluation is to verify that the installed FOV meets the specifications.

b **Requirements:** The basic measurement technique assumes a suitable test pattern fills the HUD total field of view. An example of a suitable test pattern is a grid made up of 7 horizontal lines intersecting 7 vertical lines. Using an odd number of lines provides a center reference.

Some cockpits use DEP positioning cues to indicate to the pilot that she is seated with her eyes at the cockpit DEP.

c **Action:** The pilot should sit at the cockpit DEP and view the test pattern. This test is often performed in conjunction with the fit and reach test.

d **Data:** The data consists of the pilot's subjective observation that she can see the entire test pattern.

e **Performance criteria:** The pilot shall be able to view the entire test pattern.

f **Safety considerations**: There are no particular hazards associated with this ground test.

Egress Testing

This test should be performed on new aircraft designs and one modifications that intrude into the cockpit and may hinder egress. These tests should always be performed on overhead-mount head-up displays (HUDs) or on helmet-mounted displays (HMDs).

a **Objectives**: This test verifies that the installation of the HUD/HMD does not adversely affect the flightcrews' emergency egress.

b **Requirements**: The system shall be installed in a cockpit. A simulator cockpit or a mockup is acceptable. For HMD evaluations, all mechanical inter-connections should be installed. It is not necessary for power to be supplied, although the power connections should be in place. This test should be repeated with at least three pilots ranging from fifth percentile female to ninety-fifth percentile male in size.

c **Action**: The pilot shall occupy the seat with all restraints fastened. All normal flying equipment shall be worn, including parachutes (if appropriate for the aircraft) and chemical/biological protection clothing (for military aircraft). The pilots shall perform a ground egress using normal procedures (as documented in either the CSDD or the aircraft flight manual).

Where appropriate, the pilots should perform a simulated airborne egress again using normal procedures.

d **Data**: Each pilot shall rate the equipment/interface as *satisfactory*, *acceptable*, or *unacceptable*.

e **Performance criteria**:
   i *Desired performance*: The consensus rating is satisfactory.
   ii *Adequate performance*: The consensus rating is acceptable.

f **Safety considerations**: There are no particular safety considerations.

Form and Fit Tests

These tests are required to ensure that a population of pilots covering the desired anthropometric range can be accommodated in the cockpit, have satisfactory external view, and can reach cockpit controls and switches.

a **Objectives**: The first objective will verify that the installation allows the pilot, sitting with her eyes at the design eye point (DEP) to view the various displays and external scene *and* operate all primary flight controls throughout their normal range. The second objective is to confirm that she can reach all necessary cockpit controls without changing the seat position.

b **Requirements**: The system shall be installed in a cockpit. A simulator cockpit or a mockup is acceptable. For HMD evaluations, all mechanical inter-connections should be installed. It is not necessary for power to be supplied, although the power connections should be in place. This test should be repeated with at least three pilots ranging from fifth percentile female to ninety-fifth percentile male in size.

c **Action**: The pilot shall occupy the seat with all restraints fastened. All normal flying equipment shall be worn, including parachutes (if appropriate for the aircraft). Again, military assessments should include chemical/biological protection clothing.

   Each pilot shall position her seat to place her eyes at the cockpit design eye position (DEP). While maintaining this head position, she should move all primary flight controls (stick or yoke, pedals including wheel brakes, throttles or collectives) through the controls full range of motion.

   Each pilot shall activate all cockpit controls possible without changing the seat position. A list of controls should be made indicating which controls can be actuated or not actuated (as the case may be).

   Each pilot shall ensure that all flight displays (head-down or head-up) are visible from the DEP.

   For two-pilot aircraft, these tests should be repeated for both pilot seats.

   Cockpit controls or displays intended for an additional crewmember, such as a flight engineer or loadmaster, need not require actuation by a pilot during normal operations. Such controls and displays should be examined for that crew member's use.

d **Data**: A list of all cockpit displays should be used to indicate whether the pilot can view the display:
   - while seated at the DEP
   - requires leaning, but no seat or restraint adjustment
   - requires loosening of upper torso restraint
   - can not be seen while seated.

A list of all cockpit controls should be available to indicate whether the pilot can move the control throughout the full range:
- while seated at the DEP
- requires leaning, but no seat or restraint adjustment
- requires loosening of upper torso restraint
- can not be moved while seated.

Similar lists of cockpit displays and controls should be used to indicate use by a non-pilot crewmember.
- while seated at her normal seating position
- requires leaning, but no seat or restraint adjustment
- requires loosening of upper torso restraint
- can not be moved while seated at her normal seating position.

Each crewmember shall rate the equipment/interface as *satisfactory*, *acceptable*, or *unacceptable*.

e **Performance criteria**: These criteria should be met for any crewmember within the desired population, normally from fifth percentile female to ninety-fifth percentile male in size.

  i *Desired performance*:
  - All primary flight controls can be moved throughout their full range while the pilot is seated at the DEP.
  - All cockpit controls can be moved throughout their full range while seated at the DEP or by leaning from the normal seating position without loosening of the pilot's upper torso restraint. If loosening of the upper torso restraint by a non-pilot crewmember is required, the consensus rating shall be *satisfactory*. (Note: For multicrew aircraft, this criteria is met if any crewmember can move the control. A non-pilot crewmember may loosen his upper torso restraints.)
  - No control activation requires either pilot to adjust her restraint.
  - The consensus rating is *satisfactory*.

  ii *Adequate performance*:
  - All primary flight controls can be moved throughout their full range while seated at the DEP.

- All normal cockpit controls can be moved throughout their full range while seated at the DEP, by leaning from the normal seating position. If loosening of the upper torso restraint is required, the consensus rating shall be *acceptable*. (Note: For two pilot aircraft, this criteria is met if any crewmember can move the control.)
- Emergency cockpit controls, such as alternate landing gear extension, etc., may require a crewmember to adjust or leave her seat. If a pilot must leave her seat or unfasten her restraint to activate an emergency cockpit control, the consensus rating shall be *acceptable*.
- The consensus rating is *satisfactory* or *acceptable*.

f **Safety considerations**: There are no particular safety considerations.

Symbol List Testing

The symbol list testing it is a verification that the system operates as intended and meets the description in the Crew Station Design Document (CSDD), drawings, and other design documents.

a **Objectives**: This test verifies that the various display formats can be displayed and the display modes, declutter, and occlusion logic operate according to the display specification.

b **Requirements**: The system shall be fully functional with all system inputs available. Test signals may be used vice normal aircraft signals.

c **Action**: The system shall be operated using normal procedures. Each system operating mode shall be exercised in turn verifying that all symbols appear as documented in the CSDD. Verify that text appears as documented in the CSDD.

   Within each mode, the declutter control shall be cycled through all declutter options. Verify that all symbols appear as documented in the CSDD. A sample symbol data sheet is shown in Figure 10.1.

   Within each mode, the input to the display shall be varied (with test signals or with actual aircraft data) to move each symbol throughout its normal range of motion. The occlusion logic shall be checked to ensure that proper occlusion occurs as one symbol passes near another. Verify that the occlusion logic agrees with the description in the CSDD.

   In particular, the attitude reference shall be rotated through ~90 deg in pitch, ~180 deg in roll, and 360 deg in yaw. Verify that the horizon reference complies with specifications and with the description in the CSDD.

Within each mode, verify proper mode annunciation as documented in the CSDD.

Symbol 7	Climb-Dive Marker	
Purpose	Provides flight control reference	
Data Source(s)	1 Air Data Computer 2 Vertical Gyro 3 4 5	
Update rate	TBD	
Modes/Declutter	(X) VISUAL (X) INSTRUMENT (X) LANDING (X) UNUSUAL ATTITUDE Other: see comments	No declutter No declutter No declutter No declutter
Coordinates	(X) Pitch/Roll Referenced (_) Conformal (_) Screen Fixed (_) Other	Location:
Occlusion Priority	1	
Signal Processing	See equations in Section 2	
Symbol Dynamics	TBD	
Physical Description	Winged/tailed circle	Drawing 0600-514
Effects if Invalid	Remove CDM Add Pitch Maker in Visual Mode	
Other Comments		

Figure 10.1 Sample Symbol Data Sheet
Source: Newman, Schwarz, Greeley, and Ellis (157)

Simulate input data failure (by producing an *INVALID* flag or by removing signals). Verify that each symbol's behavior agrees with the CSDD description. Verify proper failure annunciation.

Within each mode, provide input signals to simulate appropriate warnings. Verify that the display complies with the specifications and with the description in the CSDD.

If multiple input sources are possible (such as the use of gyro 1 or gyro 2), repeat the above steps with each selectable input.

d **Data**: All data should be hand recorded. Recorded data or recorded display images may be used to supplement the hand recorded data.

e **Performance criteria**: The system is acceptable if the verifications match the CSDD.

f **Safety considerations**: There are no particular safety considerations.

Flight Tests

These are preliminary tests designed to ensure that the cockpit systems meet their specifications and requirements. Mode and function testing should be performed in simulator cockpits prior to conducting simulator evaluations. Appropriate calibrations should be performed in the simulator as well.

Mode and Function Evaluation

The mode and function evaluation is not an evaluation of the suitability of the cockpit display. Rather it is a verification that the system operates as intended and meets the description in the Crew Station Design Document (CSDD), drawings, and other design documents. In the past, systems have reached flight test (indeed have reached operations) differing from the design documents. The A-7 HUD is an example. The motion of the angle-of-attack bracket differs from the description in the design specification.

a **Objectives**: The objective of this evaluation is to determine that all symbology is displayed as intended during appropriate maneuvers.

b **Requirements**: The display should be fully functional.

c **Action**: The system shall be operated using normal procedures.

Each system operating mode shall be exercised in turn verifying that all symbols appear as documented in the CSDD. Verify that text appears as documented in the CSDD.

Within each mode, the declutter control shall be cycled through all declutter options. Verify that all symbols appear as documented in the CSDD.

Within each mode, the input to the display shall be varied (by flying the aircraft within normal limits) to move each symbol throughout its normal range of motion. The occlusion logic shall be checked to ensure that proper occlusion occurs as one symbol passes near another. Verify that the occlusion logic agrees with the description in the CSDD.

In particular, the attitude reference shall be rotated through normal limits in pitch and roll, and 360 deg in yaw. Verify that the horizon reference complies with the specifications and with the description in the CSDD.

Within each mode, verify proper mode annunciation as documented in the CSDD.

To the extent permitted by safety considerations or equipment limitations, remove input data signals. Verify that each symbol's behavior agrees with the CSDD description. Verify proper failure annunciation.

If multiple input sources are possible (such as the use of number 1 or number 2 gyros), repeat the above steps with each selectable input.

Within each mode, provide input signals to simulate appropriate warnings. Verify that the display complies with the specifications and with the description in the CSDD.

d **Data**: All data should be hand recorded. Recorded data or recorded display images may be used to supplement the hand recorded data.

e **Performance criteria**: The system is acceptable if the verifications match the CSDD.

f **Safety considerations**: There are no particular safety considerations.

Calibration of Hover Vector

Airspeed and groundspeed calibrations should have been performed prior to conducting the display and related systems flight test. Techniques for such calibrations (for example, pitot-static calibration) are available in the literature.(*158*)

The helicopter hover vector is novel enough to warrant outlining the appropriate technique.

a **Objectives**: The objective of this evaluation is to obtain the displayed ground speed on the hover vector versus actual groundspeed calibration.

b **Requirements**: The hover, nap-of-the-earth (NOE), and transition modes should be fully functional.

c **Action**: The helicopter should be flown at known groundspeeds and the length of the hover vector (or other groundspeed indication) measured. This should be repeated for side and rearward flight in 45 deg increments. Normally, formation with a ground vehicle is used to establish known groundspeeds.

For simulator calibrations, the value of the groundspeed variable in the math model can be used.

d **Data**: Indicated and actual groundspeed values should be recorded simultaneously.

e **Performance criteria**: The indicated groundspeed should be within the specified tolerance of the actual groundspeed.

f **Safety considerations**: The standard low-level safety considerations apply. Two-way radio communication between aircraft and automobile should be available.

Calibration of Radar Altimeter

A radar altimeter calibration should have been performed prior to conducting the display and related systems flight test. Techniques for such calibrations (for example, pitot-static calibration) are available in the literature.(*158*)

a **Objectives**: The objective of this test is to calibrate the radar altimeter. It can also be used to obtain pitot-static (position error) calibrations.

b **Requirements**: Normally the calibration is carried out over a runway which proves a smooth surface for radar altimeter readings and a well-defined centerline. An observer station, usually an elevated tower, a known distance from the centerline is also required. The tower should be equipped with a theodolite for measuring the vertical angle as the aircraft flies by.

The height of the tower and the distance to the runway centerline must be known exactly. Figure 10.2 shows the geometry.

c **Action**: The aircraft should be flown down the runway tracking the centerline. An observer should determine the elevation angle using the theodolite. LIDAR tracking can be used as a substitute for theodolite data. Normally, differential GPS data will not be accurate enough for this test.

For radar altimeter calibrations, a runway or similar smooth, level surface is necessary. Flight at a single airspeed will suffice. Data at several altitudes should be obtained.

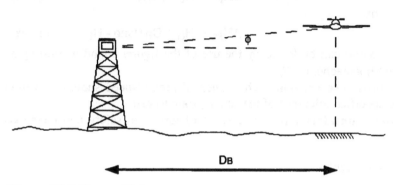

Ground Observer records φ.
Also records HG and TA for
pitot-static calibration

Pilot records indicated altitude.
Also records IAS for pitot-static
calibration

D_B

Figure 10.2 Tower Flyby

For pitot-static altimeter calibrations, the condition of the surface is less important. Barometric calibrations require flight at airspeeds throughout the speed range of the aircraft, although a single altitude is sufficient. The airplane should be flown out of ground effect — at least a wingspan above the ground.

The airplane must be flown accurately down the course line to ensure that D_B is accurate. The flight conditions do not have to be stabilized for the data to be valid.

d   **Data**: Indicated altitude values should be recorded simultaneously with tracking elevation data from the theodolite. Indicated airspeed data should be recorded for pitot-static calibrations.

From the angle recorded by the theodolite, the flight test engineer can calculate the difference in elevation between the ground station and the airplane. This is equal to the calibrated altitude of the airplane.

$$H_C = H_G + D_B * \tan\phi \qquad (10.2)$$

Normally theodolite is not corrected for non-standard temperatures since this usually results in errors of the order of one foot. If it is desired to make this correction, the following equation should be used:

$$H_C = H_G + (T_S/T_A)^*D_B^*\tan\emptyset \qquad (10.3)$$

where T_A is the measured temperature at the ground station.

The indicated altitude is simply the reading in the airplane.

Position error correction: The position error, ΔH_{PC}, is found from Equation (10.4)

$$\Delta H_{PC} = H_C - (H_I)_T \qquad (10.4)$$

or

$$\Delta H_{PC} = H_G + D_B^*\tan\emptyset - H_I \qquad (10.4a)$$

ΔV_{PC} can be found by the use of the figures found in Herrington and Shoemacher.(*158*)

e **Performance criteria**: The indicated groundspeed should be within the specified tolerance of the actual groundspeed.

f **Safety considerations**: The standard low-level safety considerations apply.

Sensor Calibrations

Sensor calibrations obviously depend on the sensor characteristics. Many of these cannot be discussed in the open literature. As a general example, this section outlines the calibration and inflight verification of weather radar systems.

a **Objectives**: The objective is to obtain sensor calibrations and verify proper operation.

b **Requirements**: The sensor and appropriate display should be fully functional.

c **Action**: The following procedure is taken from AC-23-8 (*62*).

  i *bearing accuracy*: The indicated bearing of objects shown on the display should be within +10° of their actual relative bearing. Verify that as airplane turns to right or left of target, the indicated display moves in the opposite direction. Fly under conditions which allow visual identification of a target, such as an island, a river, or a lake, at a range of approximately eighty percent of the maximum range of the radar. When flying toward the target, select a course that will pass over a reference point from which the bearing to the target is known. When flying a course from the reference point to the target, determine the error in displayed bearing to the target on all range settings. Change heading in increments of 10° and determine the error in the displayed bearing to the target.

ii *distance of operation*: The radar should be capable of displaying distinct and identifiable targets throughout the angular range of the display and at approximately eighty percent of the maximum range.

iii *beam tilting*: The radar antenna should be installed so that its beam is adjustable to any position between 10° above and below the plane of rotation of the antenna. Tilt calibration should be verified.

iv *contour display*: If heavy cloud formations or rainstorms are reported within a reasonable distance from the test base, select the contour display mode. The radar should differentiate between heavy and light precipitation. In the absence of the above weather conditions, determine the effectiveness of the contour display function by switching from normal to contour display while observing large objects of varying brightness on the indicator. The brightest object should become the darkest when switching from normal to contour mode.

v *antenna stabilization*: While in level flight at 10,000 feet or higher, adjust the tilt approximately 2 - 3° above the point where ground return was eliminated. Roll right and left approximately 15° then pitch down approximately 10° (or within design limits). No ground return should be present.

vi *ground mapping*: Fly over areas containing large, easily identifiable landmarks such as rivers, towns, islands, coastlines, etc. Compare the form of these objects on the indicator with their actual shape as visually observed from the cockpit.

vii *mutual interference*: Determine that no objectionable interference is present on the radar indicator from any electrical or radio/navigational equipment when operating and that the radar installation does not interfere with the operation of any of the airplane's radio/navigational systems.

viii *electromagnetic compatibility (EMC)*: With all systems operating in flight, verify, by observation, that no adverse effects are present in the required flight systems.

ix *light conditions*: The display should be evaluated during all lighting conditions, including night and direct sunlight.

d **Data**: Data may be hand-recorded as indicated.

e **Performance criteria**: The indicated values should be within the specified tolerance of the actual values.

f **Safety considerations**: There are no particular safety considerations.

VOR Receiver Calibrations

The following calibrations are taken from the FAA flight test advisory circulars.(*56, 57, 61, 62*) They are included here as examples of the types of calibrations necessary for civil certification.

a **Objectives**: The objective is to ensure the proper operation of the aircraft navigation receiver(s).

b **Requirements**: The VOR receiver and appropriate display modes should be fully functional. The receiver should be calibrated within the previous thirty days.

These flight tests may be reduced if adequate antenna radiation pattern studies have been made and these studies show the patterns to be without significant holes (with the airplane configurations used in flight; that is, flaps, landing gear, etc.). Particular note should be made in recognition that certain propeller rpm settings may cause modulation of the course deviation indication (prop modulation). This information should be made a part of the approved flight manual (AFM).

c **Action**:

i *reception*: The airborne VOR system should operate normally with warning flags out of view at all headings of the airplane (wings level) throughout the standard service volumes depicted in the *Airman's Information Manual* (AIM) up to the maximum altitude for which the airplane is certified.

ii *accuracy*: The accuracy determination should be made such that the indicated reciprocal agrees within 2°. Tests should be conducted over at least two known points on the ground such that data are obtained in each quadrant. Data should correlate with the ground calibration and in no case should the absolute error exceed +6°. There should be no excessive fluctuation in the course deviation indications.

iii *enroute reception*: Fly from a VOR facility rated for high altitude along a radial at an altitude of ninety percent of the airplane's maximum certificated altitude to the standard service volume range. The VOR warning flag should not come into view, nor should there be deterioration of the station identification signal. The course width should be 20° ± 5° (10° either side at the selected radial). The tests should be flown along published route segments to preclude ground station anomalies. If practical, perform an enroute segment on a Doppler VOR station to verify the compatibility of the airborne unit. Large errors have been found when incompatibility exists.

iv *low angle reception:* Perform a 360° right and left turns at a bank angle of at least 10° at an altitude just above the lower edge of the standard service volume and at the maximum service volume distance. Signal dropout should not occur as evidenced by the warning flag appearance. Dropouts that are relieved by a reduction of bank angle at the same relative heading to the station are satisfactory. The VOR identification should be satisfactory during the left and right turns.

v *high angle reception:* Repeat the turns described in (iii) above, but at a distance of 50-70 nm (20-30 nm for airplanes not to be operated above 18,000 feet) from the VOR facility and at an altitude of at least ninety percent of the maximum certificated altitude of the airplane.

vi *enroute station passage:* Verify that the to-from indicator correctly changes as the airplane passes through the cone of confusion above a VOR facility.

vii *VOR approach:* Conduct VOR approaches with gear and flaps down. With the facility 12-15 nm behind the airplane, use sufficient maneuvering in the approach to ensure the signal reception is maintained during beam tracking.

viii *electromagnetic compatibility (EMC):* With all systems operating in flight, verify, by observation, that no adverse effects are present in the required flight systems.

d **Data:** Data may be hand-recorded as indicated.

e **Performance criteria:** The indicated values should be within the specified tolerance of the actual values.

f **Safety considerations:** There are no particular safety considerations.

Localizer Calibrations

The following calibrations are taken from the FAA flight test advisory circulars.(*56, 57, 61, 62*) They are included here as examples of the types of calibrations necessary for civil certification.

a **Objectives**: The objective is to ensure the proper operation of the aircraft navigation receiver(s).

b **Requirements**: The ILS receiver and appropriate display modes should be fully functional.

   Flight test requirements should be modified to allow for adequate antenna radiation pattern measurements (*vide supra*).

c **Action**: The specific action depends on the system/sensor involved.

   i *signal strength*: The signal input to the receiver, presented by the antenna system, should be of sufficient strength to keep the malfunction indicator out of view when the airplane is in the approach configuration (landing gear extended and approach flaps) and within the normal limits of localizer coverage shown in the Airman's Information Manual (AIM). This signal should be received for 360° of airplane heading at all bank angles up to 10° left or right at all normal pitch attitudes and at an altitude of approximately 2000 feet.

   ii *bank angles*: Satisfactory results should also be obtained at bank angles up to 30° when the airplane heading is within 60° of the inbound localizer course. Satisfactory results should result with bank angles up to 15° on headings from 60° to 90° of the localizer inbound course and up to 10° bank angle on headings from 90° to 180° from the localizer inbound course.

   iii *course deviation indicator (CDI)*: The deviation indicator should properly direct the airplane back to course when the airplane is right or left of course.

   iv *station identification*: The station identification signal should be of adequate strength and sufficiently free from interference to provide positive station identification, and voice signals should be intelligible with all electric equipment operating and pulse equipment transmitting.

   v *localizer intercept*: In the approach configuration and at a distance of at least 18 nm from the localizer facility, fly toward the localizer front course, inbound, at an angle of at least 50°. Perform this maneuver from both left and right of the localizer beam. No flags should appear during the time the deviation indicator moves from full deflection to on course.

vi *localizer tracking*: While flying the localizer inbound and not more than 5 miles before reaching the outer marker, change the heading of the airplane to obtain full needle deflection. Then fly the airplane to establish localizer on course operation. The localizer deviation indicators should direct the airplane to the localizer on course. Perform this maneuver with both a left and a right needle deflection. Continue tracking the localizer until over the transmitter. Acceptable front course and back course approaches should be conducted to 200 feet or published minimums.

vii *electromagnetic compatibility (EMC)*: With all systems operating in flight, verify, by observation, that no adverse effects are present in the required flight system.

d **Data**: Data may be hand-recorded as indicated.

e **Performance criteria**: The indicated values should be within the specified tolerance of the actual values.

f **Safety considerations**: There are no particular safety considerations.

Glideslope Calibrations

The following calibrations are taken from the FAA flight test advisory circulars.(*56, 57, 61, 62*) They are included here as examples of the types of calibrations necessary for civil certification.

a **Objectives**: The objective is to ensure the proper operation of the aircraft navigation receiver(s).

b **Requirements**: The ILS receiver and appropriate display modes should be fully functional.

c **Action**: The specific action depends on the system/sensor involved.

i *signal strength*: The signal input to the receiver should be of sufficient strength to keep the warning flags out of view at all distances to 10 nm from the facility. This performance should be demonstrated at all airplane headings between 30° right and left of the localizer course. The deviation indicator should properly direct the airplane back to path when the airplane is above or below path. Interference with the navigation operation, within 10 nm of the facility, should not occur with all airplane equipment operating and all pulse equipment transmitting. There should be no interference with other equipment as a result of glideslope operation.

ii *glideslope tracking*: While tracking the glideslope, maneuver the airplane through normal pitch and roll attitudes. The glideslope deviation indicator should show proper operation with no flags. Acceptable approaches to 200 feet or less above threshold should be conducted.

iii *electromagnetic compatibility (EMC)*: With all systems operating in flight, verify, by observation, that no adverse effects are present in the required flight systems.

d **Data**: Data may be hand-recorded as indicated.

e **Performance criteria**: The indicated values should be within the specified tolerance of the actual values.

f **Safety considerations**: There are no particular safety considerations.

Automatic Direction Finding Radio Calibrations

The following calibrations are taken from the FAA flight test advisory circulars.(*56, 57, 61, 62*) They are included here as examples of the types of calibrations necessary for civil certification.

Automatic Direction Finder (ADF) calibrations are also included since many design engineers and flight test personnel may not be familiar with ADF navigation, believing it to be obsolete. There are a significant number of ADF facilities and procedures, particularly in the more remote regions of the world.

a **Objectives**: The objective is to ensure the proper operation of the aircraft navigation receiver(s).

b **Requirements**: The ADF receiver and appropriate display modes should be fully functional.

c **Action**

i *range and accuracy*: The ADF system installed in the airplane should provide operation with errors not exceeding 5°, and the aural signal should be clearly audible up to the distance listed for any one of the following types of radio beacons:

- 75 nm from an HH facility.
- 50 nm from an H facility. Caution - service ranges of individual facilities may be less than 50 nm.
- 25 nm from an MH facility.

- 15 nm from a compass locator.

ii *needle reversal*: The ADF indicator needle should make only one 180° reversal when the airplane flies over a radio beacon. This test should be made with and without the landing gear extended.

iii *indicator response*: When switching stations with relative bearings differing by 180° +5°, the indicator should indicate the new bearing within +5° in not more than 10 seconds.

iv *antenna mutual interaction*: For dual installations, there should not be excessive coupling between the antennas.

Tune in a number of radio beacons spaced throughout the 190 - 535 KHz range and located at distances near the maximum range for the beacon. The identification signals should be understandable and the ADF should indicate the approximate direction to the stations. Beginning at a distance of at least 15 nm from a compass locator in the approach configuration (landing gear extended, approach flaps), fly inbound on the localizer front course and make a normal ILS approach. Evaluate the aural identification signal for strength and clarity and the ADF for proper performance with the receiver in the ADF mode. All electrical equipment on the airplane should be operating and all pulse equipment should be transmitting. Fly over a ground or appropriately established checkpoint with relative bearings to the facility of 0°, 45°, 90°, 135°, 180°, 225°, 270°, and 315°. The indicated bearings to the station should correlate within 5°. The effects of the landing gear on bearing accuracy should be determined. (A calibration placard should be provided, if appropriate.)

v *needle reversal*: Fly the airplane over an H, HH, or compass locator facility at an altitude 1000 to 2000 feet above ground level. Partial reversals which lead or lag the main reversal are permissible.

vi *indicator response*: With the ADF indicating station dead ahead, switch to a station having a relative bearing of 175°. The indicator should indicate within +5° of the bearing in not more than 10 seconds.

vii *antenna mutual interaction*: If the ADF installation being tested is dual, check for coupling between the antenna by using the following procedure. With number 1 ADF receiver tuned to a station near the low end of the ADF band, tune the number 2 receiver slowly throughout the frequency range of all bands and determine whether the number 1 ADF indicator is adversely affected. Repeat with the number 1 ADF receiver tuned to a station near the high end of the ADF band.

viii *electromagnetic compatibility (EMC)*: With all systems operating in flight, verify, by observation, that no adverse effects are present in the required flight systems.

d **Data**: Data may be hand-recorded as indicated.

e **Performance criteria**: The indicated values should be within the specified tolerance of the actual values.

f **Safety considerations**: There are no particular safety considerations.

Distance Measuring Equipment Calibrations

The following calibrations are taken from the FAA flight test advisory circulars.(*56, 57, 61, 62*) They are included here as examples of the types of calibrations necessary for civil certification.

a **Objectives**: The objective is to ensure the proper operation of the aircraft navigation receiver(s).

b **Requirements**: The Distance Measuring Equipment (DME) receiver and appropriate display modes should be fully functional.

c **Action**:

i *tracking performance*: The DME system should continue to track without dropouts when the airplane is maneuvered throughout the airspace within the standard service volume of the VORTAC/DME station and at altitudes above the lower edge of the standard service volume to the maximum operating altitude. This tracking standard should be met with the airplane:

- In cruise configuration.
- at bank angle up to 10°.
- climbing and descending at normal maximum climb and descent attitude.
- orbiting a DME facility.
- provide clearly readable identification of the DME facility.

ii *climb and maximum distance*: Determine that there is no mutual interference between the DME system and other equipment aboard the airplane. Beginning at a distance of at least 10 nm from a DME facility and at an altitude of 2000 feet above the DME facility, fly the airplane on a heading so that the airplane will pass over the facility. At a distance of 5 - 10 nm beyond the DME facility, operate the airplane at its normal maximum climb attitude up to ninety percent of the maximum operating altitude, maintaining the airplane on a station radial (within 5°). The DME should continue to track with no unlocks to the range of the standard service volume.

iii *long range reception*: Perform two 360° turns, one to the right and one to the left, at a bank angle of at least 10° at the maximum service volume distance of the DME facility and at an altitude of at least ninety percent of the maximum operating altitude. Unlocks may occur and are acceptable if they do not interfere with the intended flight path of the airplane or are relieved by a reduction of bank angle at the same relative heading to the station.

iv *high angle reception*: Repeat the flight pattern and observations of (iii) above at a distance of 50-70 nm (20-30 nm for airplanes not to be operated above 18,000 feet) from the DME facility and at an altitude of at least ninety percent of the maximum operating altitude.

v *penetration*: From ninety percent of the maximum operating altitude, perform a letdown directly toward the ground station using normal maximum rate of descent procedures to a DME facility so as to reach an altitude of 5000 feet above the DME facility 5-10 nm before reaching the DME facility. The DME should continue to track during the maneuver with no unlocks.

vi *orbiting*: At an altitude of 2000 feet above the terrain, at holding pattern speeds appropriate for the type of airplane and with the landing gear extended, fly at least 15° sectors of left and right 35 nm orbital patterns around the DME facility. The DME should continue to track with no more than one unlock, not to exceed one search cycle, in any 5 miles of orbited flight.

vii *approach*: Make a normal approach at an actual or simulated field with a DME. The DME should track without an unlock (station passage expected).

viii *DME hold*: With the DME tracking, activate the DME hold function. Change the channel selector to a localizer frequency. The DME should continue to track on the original station.

ix *electromagnetic compatibility (EMC)*: With all systems operating in flight, verify, by observation, that no adverse effects are present in the required flight systems.

d **Data**: Data may be hand-recorded as indicated.

e **Performance criteria**: The indicated values should be within the specified tolerance of the actual values.

f **Safety considerations**: There are no particular safety considerations.

Speech Intelligibility testing

This test is a standard evaluation of audio systems, particularly radio communications or intercom systems. In the past this test has been required, particularly for noisy aircraft. It has been omitted in some aircraft, but justifying omission often requires more effort than simply conducting the test.

The words spoken should be chosen to avoid context cues helping the evaluators to interpret them. Consequently, standard word lists are used.(*159*)

a **Objectives**: This test verifies that the audio system allows the flight crew to understand speech communications. A secondary objective will verify that audio warning using synthetic speech is intelligible.

b **Requirements**: The system shall be installed in an aircraft cockpit. A simulator cockpit would be used for preliminary testing of audio warning systems.

c **Action**: All crewmembers shall occupy the normal positions using a head-set or helmet. For use by a the mobile crewmember, such as a loadmaster, that crewmember will move about as if performing her normal duties.

Transport crews should perform the test using normal head-sets including oxygen masks. The tests should be repeated using cabin speakers.

For military aircraft, the tests should be repeated wearing chemical/biological protection clothing.

The test is conducted with a speaker reading from a list of test words. The evaluator interprets the word.

Communications testing must be conducted inflight at typical station-to-station ranges. Tests conducted beyond 60 nm from the facility should ensure that the communications facility is an enroute facility. Within 60 nm, terminal communications facilities may be used.

i *long range reception*: Perform two 360° turns, one to the right and one to the left, at a bank angle of at least 10° at 150 nm at an altitude of at least ninety percent of the maximum operating altitude. For airplanes with a maximum operating altitude less than 24000 ft, perform these tests at a range of 60 nm.

ii *high angle reception*: Repeat the flight pattern and observations of (i) above at a distance of 50-70 nm (20-30 nm for airplanes not to be operated above 18,000 feet) from the communications facility and at an altitude of at least ninety percent of the maximum operating altitude.

iii *ground reception*: Conduct the tests while on the ground to ensure satisfactory operations

d **Data**: Some tests (*160*) require the evaluator to write the word; the modified rhyme test (MRT) give an answer sheet with a number of words to pick from. The choice of words might be "sun", "sub", "sum", "sud", "sup", or "sung". The evaluator need simply circle or check the word. Between thirty and fifty words are normally read for each test.

e **Performance criteria**:
i *Desired performance*: Ninety-five percent of the words should be correctly identified.
ii *Adequate performance*: Ninety percent of the words should be correctly identified.

f **Safety considerations**: There are no particular safety considerations.

Display Dynamic Response

These are preliminary tests designed to obtain display dynamic response data for diagnostic purposes. These dynamic response tests need not be performed in the simulator unless there is evidence of a lack of simulator fidelity. Such tests are primarily for diagnostic purposes.

Following these open-loop dynamic testing, the first closed loop test should be the aggressive air-to-ground visual tracking (page 154 for rotary-wing aircraft and page 187 for fixed-wing).

Initial Dynamics Testing: Single-axis Control Steps or Pulses

These tests are diagnostic in nature. The purpose is to obtain display response data as a result of step and pulse control inputs.

a **Objectives**: The objective of this evaluation is to obtain display response data to be used in system diagnostics.

b **Requirements**: The display should be fully functional.

c **Action**: The aircraft/display shall be tested in accordance with the methods described in section 3 of ADS-33 (*137*) or in MIL-F-8785.(*138*) For HMD evaluations, the pilot shall maintain his head in a fixed direction during these tests.

d **Data**: Aircraft state data shall be recorded as time histories. Display data bus input data time histories shall be recorded electronically. Recording display video images should be recorded, if possible.

e **Performance criteria**: There are no pass/fail requirements. These tests are performed to obtain response data for system diagnostics or to develop simulator models.

f **Safety considerations**: The standard flight test safety considerations apply.

Initial Dynamics Testing: Step Inputs to Head-Tracker (in Flight)

These tests are diagnostic in nature. The purpose is to obtain head-tracker response data as a result of step and pulse control inputs.(*136*) These tests only apply to head-tracked systems.

a **Objectives**: The objective of this evaluation is to obtain the interaction of HMD dynamics with aircraft dynamics to be used in system diagnostics.

b **Requirements**: The HMD should be fully functional.

c **Action**: With the aircraft in straight-and-level, unaccelerated flight, the pilot should rapidly move her line-of-sight (LOS) from straight ahead to a point off the right wing (at least ninety-degrees). She should look as far back as practical and move her LOS rapidly to straight ahead. This should be repeated to the left.

Again, with the aircraft in straight-and-level, unaccelerated flight, the pilot should rapidly move her line-of-sight (LOS) from straight ahead to a point as far up as practical. After a brief pause, she should shift her LOS rapidly to straight ahead. This should be repeated in a downward direction.

These tests should be repeated while flying in left and right steep turns (with bank angles appropriate to the type of aircraft).

d **Data**: Aircraft state data time histories shall be recorded electronically. Display data bus input data time histories shall be recorded electronically. Recording display video images should be recorded, if possible.

e **Performance criteria**: There are no pass/fail requirements. These tests are performed to obtain response data for system diagnostics or to develop simulator models.

f **Safety considerations**: The standard flight test safety considerations apply. Rapid head motion can contribute to motion sickness or spatial disorientation. Suitable precautions should be taken, such as having another pilot fly while the pilot not-flying (PNF) performs the head step inputs.

Initial Dynamics Testing: Step Inputs to Head-Tracker (on Ground)

These tests are diagnostic in nature. The purpose is to obtain head-tracker response data as a result of step and pulse control inputs. These tests only apply to head-tracked systems.

a **Objectives**: The objective of this evaluation is to obtain HMD and head-tracker response data separate from aircraft dynamics to be used in system diagnostics.

b **Requirements**: The HMD should be fully functional.

c **Action**: With the aircraft static on the ground, the pilot should rapidly move her line-of-sight (LOS) from straight ahead to a point off the right wing (at least ninety degrees). She should look as far back as practical and move her LOS rapidly to straight ahead. This should be repeated to the left.

Again, with the aircraft static on the ground, the pilot should rapidly move her line-of-sight (LOS) from straight ahead to a point as far up as practical. After a brief pause, she should move her LOS rapidly to straight ahead. This should be repeated in a downward direction.

d **Data**: Aircraft state data shall be recorded. Display data bus input data time histories shall be recorded electronically. Recording display video images should be recorded, if possible.

e **Performance criteria**: There are no pass/fail requirements. These tests are performed to obtain response data for system diagnostics or to develop simulator models.

f **Safety considerations**: There are no particular safety considerations.

Collateral Tests

During the course of the display evaluation, most of the specific evaluations will be preplanned. However, the evaluation pilots should be observant for display and cockpit related problems and report them. In past programs,(9) evaluation pilots have used standard questionnaires asking for specific problems noted during the flight.

Display Readability

The standard questionnaire should inquire about optical deficiencies. Optical deficiencies include
- Flicker,
- Excessive brightness variation across the FOV,
- Symbol distortion,
- Distortion of real world objects (HUDs and HMDs),
- Legibility of symbols,
- Secondary symbol images,
- Reflections of real world images and lights,
- Visual disparity (HUDs and HMDs),
- Interference with view of real world cues (HUDs and HMDs).

Figure 10.3 shows a standard data card to be completed following the flight.

Display Dynamics

The standard questionnaire should inquire about display dynamics deficiencies. Display dynamics deficiencies include
- Jitter,
- Excessive lead or lag,
- Excessive noise,
- Mutual interference of symbols.

Figure 10.3 shows a standard data card to be completed following the flight.

Display Utility

The standard questionnaire should inquire about difficulties encountered with system controls, modes, and functions, during the course of the flight. Figure 10.4 shows a standard data card to be completed following the flight.

DISPLAY READABILITY DATA CARD		Page 1
PILOT NAME:	SORTIE/DATE:	
ALTITUDE:	FLIGHT TASK:	
BACKGROUND SCENE:	WEATHER:	DAY(_)/NIGHT(_)

Optical/visual problems noted:?

Problem noted	Comments/Describe
(_) Flicker	
(_) Brightness Variations Across FOV	
(_) Symbol Distortion	
(_) Symbol Illegibility	
(_) Secondary Symbol Images	
(_) Stray Reflections	
(_) Visual Disparity	
(_) Difficulty in Focusing	
(_) Double Vision	
(_) Interference with Real World Cues	

Dynamics/Motion problems noted:?

Problem noted	Comments/Describe
(_) Jitter	
(_) Excessive Lead or Lag	
(_) Excessive Noise	
(_) Mutual Interference of Symbols	
(_) Symbol motion not Natural	

Any other problems, comments, or observations?

Figure 10.3 Display Readability Data Card

DISPLAY UTILITY DATA CARD	Page 2

PILOT NAME:	SORTIE/DATE:

Was the HUD suitable for the flight tasks flown?
Comments

Were the following functions used?
Used Mode   Comments
(_) ENROUTE
(_) INSTRUMENT
(_) APPROACH
(_) DECLUTTER

Were there any problems with detecting other traffic caused by the HUD?

Were the flight manual procedures clear and appropriate?

Any other problems, comments, or observations?

Figure 10.4 Display Utility Data Card

11 Rotary-Wing Mission Task Elements

Aggressive Mission Task Elements

The following are aggressive mission task elements (MTEs) for rotary-wing aircraft. In most cases, they were adapted from other test and evaluation documents and modified to examine display characteristics and display/aircraft interactions, rather than aircraft characteristics alone.

Acceleration/Deceleration

Acceleration/deceleration (Accel/decel) is an aggressive maneuver consisting of a rapid acceleration to a desired speed followed by a rapid deceleration back to a hover. It is used in ADS-33 (*137*) to evaluate the transition from hover to full translational lift.

a **Objectives**: The objectives are threefold. The first objective is to check pilot's ability to use the display to aggressively control the aircraft during aggressive maneuvering in the longitudinal axis.

    The second objective checks for aircraft/display harmony during these maneuvers. Actually, the test pilot should look for undesirable aircraft/display interactions during rapid transitions to-and-from hovering flight. The final objective evaluates the acceptability of the transition symbology during aggressive transitions.

b **Requirements**: The test course shall consist of a reference line between the start and end points of the maneuver and markers indicating the start and end points. The distance between these points will be a function of the aircraft's performance and should be determined based on trial runs consisting of accelerations to the target airspeed and decelerations to hover. If the display allows, this evaluation should be flown without using these test course artifacts, to determine if the display provides sufficient cues to the maneuver.

152

c **Action**: From a stabilized hover, rapidly increase power to approximately maximum, accelerate in the forward direction, and maintain constant altitude with cyclic. Hold collective constant during the acceleration to an airspeed of 50 knots. Upon reaching the target airspeed, initiate a deceleration by aggressively reducing the power and holding altitude constant with cyclic. Return to a stabilized hover.

The target power is ninety-five percent of maximum continuous power or ninety-five percent of maximum transient power that can be sustained for the required acceleration. If this power results in objectionable pitch attitudes, then use the maximum nose-down pitch attitude that is felt to be acceptable. This pitch attitude shall be considered as a limit of the Operational Flight Envelope for NOE flying for the visual conditions being evaluated.

If the acceleration is very rapid, it may be desirable to extend the course or use less power to allow for better evaluation of the display.

d **Data**: Aircraft performance data should be recorded as functions of time:

> Airspeed
> Altitude
> Pitch and bank angles
> Heading.

Aircraft position should be obtained from LIDAR or other optical tracking system. Differential GPS may be used for aircraft position as well,

Pilot ratings (DRRs and DFRs) should be hand recorded. Voice recording may be used to supplement the hand recorded data.

e **Performance criteria**:

i *Desired performance*:

- Complete the maneuver over the reference point at the end of the course. The longitudinal tolerance is +0/-0.5L, where L is the overall length of the aircraft.
- Maintain lateral track within ±10 ft.
- Maintain altitude below 50 ft.
- Maintain heading within ±10 deg.
- Achieve target power within 1.5 sec from initiation of the maneuver.

- Decrease power to full down collective within 3 sec with no significant increases in power until just before the final stabilized hover.
- Achieve a nose-up attitude during the deceleration of at least 30 deg above hover attitude.
- Rotor rpm shall remain within Operational Flight Envelope without undue pilot compensation.

ii *Adequate performance*:
- Complete the maneuver over the reference point at the end of the course. The longitudinal tolerance is +0/-L, where L is the overall length of the aircraft.
- Maintain lateral track within ±10 ft.
- Maintain altitude below 70 ft.
- Maintain heading within ±20 deg.
- Achieve target power within 3 sec from initiation of the maneuver.
- Decrease power to full down collective within 3 sec with no significant increases in power until just before the final stabilized hover.
- Achieve a nose-down attitude during the deceleration of at least 7 deg below hover attitude.
- Achieve a nose-up attitude during the deceleration of at least 10 deg above hover attitude.
- Rotor rpm shall remain within Service Flight Envelope.

f **Safety considerations**: These tests are aggressive maneuvers flown in proximity to the ground and obstructions. Suitable precautions for low-level flight should be observed.

Aggressive Air-to-Ground Visual Tracking

Air-to-ground visual tracking as an extremely useful test to determine if there are any problems with the short period dynamics of a see-through display during aggressive maneuvering.(*136*) This task should be performed on all aircraft equipped with HUDs or HMDs regardless of their intended missions. Aggressive visual tracking is inappropriate for simulator evaluation because of possible interactions between display dynamics and visual scene delays.

a **Objectives**: The objective is to check pilot-vehicle-display dynamics when the pilot is forced into tight compensatory tracking behavior.

b **Requirements**: This evaluation requires a convenient ground target.

c **Action**: The maneuver is initiated from descending flight toward the target at the desired airspeed. For all aircraft, landing reference speed, approximately a 3 deg descent, and a three mile approach leg should be used. For air-to-ground aircraft, weapon delivery airspeed, a suitable dive angle, and a suitable distance should also be used. The test aircraft is flown toward the target maintaining the aiming symbol (or aircraft reference symbol) on the ground target. At the approximate halfway point, the aimpoint should be shifted to a new aimpoint. The maneuver should be discontinued at a safe altitude.

d **Data**: Aircraft performance data should be recorded as functions of time:

> Airspeed
> Altitude
> Pitch and bank angles.

Sight picture should be recorded using visual image recording.

Pilot ratings (DRRs and DFRs) should be hand recorded. Voice recording may be used to supplement the hand recorded data.

e **Performance criteria**:

i *Desired performance*:
- Acquire new target with no tendency for overshoots.
- Maintain aimpoint within 5 mrad or within firing constrains of the weapon system to be deployed on the aircraft.
- Acquire new target with no tendency for overshoots.
- Maintain aimpoint within 5 mrad or within firing constrains of the weapon system to be deployed on the aircraft.
- No undesirable cross-coupling.

ii *Adequate performance*:
- Acquire new target with minimal tendency for overshoots.
- Maintain aimpoint within 10 mrad or within firing constrains of the weapon system to be deployed on the aircraft.
- Acquire new target with minimal tendency for overshoots.
- Maintain aimpoint within 10 mrad or within firing constrains of the weapon system to be deployed on the aircraft.
- No objectionable cross-coupling.

f **Safety considerations**: These tests are aggressive maneuvers flown in proximity to the ground and obstructions. Suitable precautions for low-level flight should be observed.

Bob-Up/Bob-Down

The bob-up/bob-down is a maneuver emulating helicopter combat tactics. It is similar to the maneuver used in ADS-33 (*137*) to check the vertical handling qualities. In display evaluation, it has the additional benefit of ascertaining that the display can allow the pilot to maintain a fixed geographic position while maneuvering vertically.

a **Objectives**: The objective checks the pilot's ability to use the display to precisely start and stop a vertical rate and to maintain a position over a predetermined location.

b **Requirements**: The test course should include markings to indicate the target location for the hover. It may be desirable to include a vertical reference (such as a telephone pole) to provide cues for the upper vertical altitude. If the display allows, this evaluation should be flown without using these test course artifacts, to determine if the display provides sufficient cues to the maneuver. Any convenient target may be used for the aimpoint.

c **Action**: From a stabilized hover at 10 ft, bob-up to a defined reference altitude between 40 and 50 ft. Stabilize at the reference altitude with the aiming reticle on the target for at least 2 sec for at least two seconds, simulating an attack on the target with guns. Bob-down to re-establish the 10 ft stabilized hover. This maneuver is to be accomplished with the wind from the most critical azimuth. If the most critical azimuth has not been defined, the maneuver shall be accomplished with a direct tailwind.

d **Data**: Aircraft performance data should be recorded as functions of time:
> Airspeed
> Radar altitude.

Heading angle, target acquisition, and tracking accuracy shall be recorded, preferably using visual image recording.

Pilot ratings (DRRs and DFRs) should be hand recorded. Voice recording may be used to supplement the hand recorded data.

e **Performance criteria**:
  i *Desired performance*:
- Maintain longitudinal and lateral position of a selected point on the aircraft within 6 ft of a selected point on the ground.
- Maintain bob-up altitude within ±3 ft.
- Stabilize bob-up heading within a tolerance based on the firing constraints of the weapon system to be deployed on the aircraft.
- Maintain final stabilized altitude within ±3 ft.
- Stabilize final aircraft heading within ± 3 deg.
- Complete the maneuver within 10 seconds.

  ii *Adequate performance*:
- Maintain longitudinal and lateral position of a selected point on the aircraft within 10 ft of a selected point on the ground.
- Maintain bob-up altitude within ±6 ft.
- Stabilize bob-up heading within ± 3 deg.
- Maintain final stabilized altitude within ±6 ft.
- Stabilize final aircraft heading within ± 6 deg.
- Complete the maneuver within 15 seconds.

f **Safety considerations**: These tests are flown in proximity to the ground and obstructions. Standard precautions for low-level flight should be observed. Particular care should be taken during the aggressive descent portion of the test.

Engine Failure

a **Objectives**: These tests are designed to verify that the pilot can recognize engine failures and take appropriate corrective action while maintaining aircraft control.

b **Requirements**: These tests shall be flown in conjunction with other evaluations being conducted in order to present the evaluation pilot with unexpected failures.

    Engine failures should be introduced during mission related tasks to determine the reaction time and responses of the pilot. Appropriate procedures for operating the aircraft following these failures must be established prior to conducting these tests. For single-engine helicopters, these tests will involve entry into an autorotation.

c **Action**: For tests involving operations in IMC, the evaluation pilot shall be hooded or otherwise prevented from using outside visual cues.

Engine failures shall be introduced during operations at unexpected times by the safety pilot.

Upon recognition of the failure, the evaluation pilot shall take appropriate action as described in the aircraft/system operating procedures. The evaluation pilot will be briefed to make a verbal call upon recognition of all system failures as a means of determining detection time.

Following recovery, the evaluation pilot shall complete displays ratings (DRR and DFR).

d **Data**: Aircraft performance data should be recorded as functions of time:

> Airspeed
> Altitude
> Pitch and bank angles
> Event marker
> Engine parameters appropriate to failure.

Pilot ratings (DRRs and DFRs) should be hand recorded. Voice recording may be used to supplement the hand recorded data.

e **Performance criteria**:

i *Desired performance*:

- Initial correct engine control input within 4 sec.
- Initial engine control input in accordance with published standards. (Published standards include aircraft-specific manuals or equivalent documents.)
- Maintain altitude within ±200 ft. (engine-out performance permitting).
- Maintain airspeed within ±5 knots.
- Maintain heading within ±10 deg.

ii *Adequate performance*:

- Initial correct engine control input within 8 sec.
- Initial engine control input in accordance with published standards. (Published standards include aircraft-specific manuals or equivalent documents.)
- Maintain altitude within ±400 ft. (engine-out performance permitting).
- Maintain airspeed within ±10 knots.
- Maintain heading within ±30 deg.

f **Safety considerations**: These tests involve simulated engine failure and shut-down. Suitable precautions for such maneuvers should be observed.

For single-engine helicopters, these tests will involve entry into an autorotation and subsequent engine-out forced landing. The classic argument of whether or not to continue the autorotation to touchdown or to recover with power is equally valid in these tests. Certainly during the handling qualities tests, autorotations to touchdown will have been performed. These flight tests will need to consider the handling of the aircraft during autorotation and determine if the extra hazard compared with the additional data warrants full-touchdown autorotations.

Flight Control System Failures

These tests are intended to evaluate the ability of the pilot to use the display to recover from flight control failures. Situation awareness and workload will be evaluated in mission-related tests. These types of failures are adapted from autopilot failure testing (56-57, 63) or stability augmentation failure testing.(161) The general method is to introduce the failure at unexpected times and determine the resultant deviation from the desired aircraft flight path.

a **Objectives**: These tests are designed to verify that the pilot can recognize critical flight control system (FCS), stability augmentation system (SAS), autopilot system, and display dynamic system failures and take appropriate corrective action while maintaining aircraft control.

b **Requirements**: These tests shall be flown in conjunction with other evaluations being conducted.

FCS, SAS, autopilot, and display system failures shall be introduced during mission related tasks to determine the reaction time and responses of the pilot. Appropriate procedures for operating the aircraft following these failures shall be established prior to conducting these tests.

Typical failures include
- FCS failures (change in gains or loss of function)
- SAS failures (hardover or loss of SAS function)
- Autopilot failure (both hardover and slow failures)
- Loss of quickening or change in quickening function
- Change in display damping.

These failures will normally be induced by the use of a special software load that will allow the flight test engineer to introduce these failures at the required point in time. The behavior of these special software functions must be verified prior to conducting these tests. In some case, these failures may be generated by pulling circuit breakers, either on the system in question on or sensors providing inputs for gain scheduling.

c **Action**:   For tests involving operations in IMC, the evaluation pilot shall be hooded or otherwise prevented from using outside visual cues. System failures shall be introduced during operations at unexpected times by the safety pilot or flight test engineer.

Upon recognition of the failure, the evaluation pilot shall take appropriate action as described in the aircraft/system operating procedures. The evaluation pilot will be briefed to make a verbal call upon recognition of all system failures as a means of determining detection time.

Autopilot testing normally requires a delay (one to three seconds) after recognition before the pilot takes corrective action. These delays depend on the degree of pilot attention. In the systems failure tests covered in this section, no such recognition time will be required since the pilot will be actively involved in flight control.

Following recovery, the evaluation pilot shall complete displays' ratings (DRR and DFR).

d **Data**: Aircraft performance data should be recorded as functions of time:

> Airspeed
> Altitude
> Pitch and bank angles
> Event marker
> System parameters appropriate to failure.

Pilot ratings (DRRs and DFRs) should be hand recorded. Voice recording may be used to supplement the hand recorded data.

e **Performance criteria**:
  i *Desired performance*:
   • Failure detection time within 2 sec of first indication. Slowly developing failures, such as autopilot softovers may require addition time. In such cases, it may be better to describe the de-

sired and adequate performance in terms of flight path deviations rather than time.

- Pilot actions in accordance with published standards. Published standards include aircraft-specific manuals or equivalent documents.
- Maintain altitude within ±200 ft.
- Maintain airspeed within ±10 knots.
- Maintain heading within ±10 deg.

ii *Adequate performance*:

- Failure detection time within 5 sec of first indication. Slowly developing failures, such as autopilot softovers may require addition time. In such cases, it may be better to describe the desired and adequate performance in terms of flight path deviations rather than time.
- Pilot actions in accordance with published standards. Published standards include aircraft-specific manuals or equivalent documents.
- Maintain altitude within ±500 ft.
- Maintain airspeed within ±20 knots.
- Maintain heading within ±20 deg.

f **Safety considerations**: These tests involve simulated flight control failures. Suitable precautions for should be observed.

Sidestep

Sidestep is an aggressive maneuver consisting of a rapid lateral acceleration followed by a rapid deceleration back to a hover. It is used in ADS-33 (*137*) to evaluate aggressive lateral maneuvering.

a **Objectives**: There are three objectives. First, to check the pilot's ability to use the display to aggressively control the aircraft during aggressive lateral maneuvering.

The second objective checks for the absence of undesirable aircraft/display interactions during rapid lateral transitions to-and-from hovering flight. The final objective evaluates the acceptability of the transition symbology during aggressive lateral maneuvers.

b **Requirements**: The test course consists of any reference line on the ground that is visible to the pilot.

c **Action**: From a stabilized hover at a selected altitude below 30 ft, initiate a rapid and aggressive lateral translation, with a bank angle of 25 deg, holding altitude with power. When the aircraft has achieved a lateral airspeed of 45 knots (or five knots below the maximum allowable lateral airspeed), immediately initiate an aggressive deceleration to hover at a constant altitude. The peak bank angle during deceleration should be at least 30 deg and should occur just before the aircraft comes to a stop. Establish and maintain a stabilized hover for 5 sec. Immediately repeat the maneuver in the opposite direction. The aircraft heading should be at right angles to the course line throughout the maneuver.

Tests in degraded visual conditions may restrict the maximum bank angle to 20°.(*147*)

The maneuver may be performed in calm winds or in a headwind of less than 10 knots.

d **Data**: Aircraft performance data should be recorded as functions of time:

> Airspeed
> Altitude
> Pitch and bank angles
> Heading.

Aircraft position should be obtained from LIDAR or other optical tracking system. Differential GPS may be used for aircraft position as well,

Pilot ratings (DRRs and DFRs) should be hand recorded. Voice recording may be used to supplement the hand recorded data.

e **Performance criteria**:

i *Desired performance*:

- Maintain the selected point on the aircraft within 10 ft of the ground reference line.
- Maintain altitude within ±10 ft.
- Maintain heading within ± 10 deg.
- Achieve at least 25 deg of bank within 1.5 sec of initiating side-step.
- Achieve at least 30 deg of bank within 1.5 sec of initiating deceleration.
- Achieve stabilized hover within 5 sec after reaching hover point.

ii *Desired performance*:
- Maintain the selected point on the aircraft within 15 ft of the ground reference line.
- Maintain altitude within ±15 ft.
- Maintain heading within ± 15 deg.
- Achieve at least 25 deg of bank within 3 sec of initiating side-step.
- Achieve at least 30 deg of bank within 3 sec of initiating deceleration.
- Achieve stabilized hover within 10 sec after reaching hover point.

f **Safety considerations**: These tests are aggressive maneuvers flown in proximity to the ground and obstructions. Suitable precautions for low-level flight should be observed.

Slalom

The slalom is a high-speed task involving maneuvering between objects placed on the ground. It is used in ADS-33 (*137*) to evaluate aircraft handling qualifies.

a **Objectives**: Check pilot's ability to use the display to aggressively maneuver during forward flight by reference to objects on the ground.

b **Requirements**: The test course consists of a series of gates marked by pylons or other markings and shown in Figure 11.1. Most runways have touchdown markings at 500 ft intervals that can be conveniently used instead of pylons.

$\leftarrow$500 ft$\rightarrow$

Figure 11.1 Slalom Test Course
Source: *ADS-33 (137)*

c **Action**: Initiate the maneuver in level unaccelerated flight along the centerline of the course. Perform a series of smooth turns at 500 ft intervals maintaining an altitude less than 100 ft. Complete the course on the centerline.

d **Data**: Aircraft performance data should be recorded as functions of time:

> Airspeed
> Altitude
> Pitch and bank angles
> Heading.

Sight picture should be recorded using visual image recording.

Aircraft position should be obtained from LIDAR or other optical tracking system. Differential GPS may be used for aircraft position as well.

Pilot ratings (DRRs and DFRs) should be hand recorded. Voice recording may be used to supplement the hand recorded data.

e **Performance criteria**:

  i *Desired performance*:
  - Maintain an airspeed of at least 60 knots throughout the course.
  - Maintain altitude within ±10 ft.

  ii *Adequate performance*:
  - Maintain an airspeed of at least 40 knots throughout the course.
  - Maintain altitude within ±20 ft.

f **Safety considerations**: These tests are aggressive maneuvers flown in proximity to the ground and obstructions. Suitable precautions for low-level flight should be observed.

Turn-to-Target

For combat helicopters, the turn-to-target is a handling qualities task *(137)* appropriate for HUDs or HMDs. During tests of HMDs, the pilot shall keep his head aligned with the aircraft longitudinal axis (i. e. turn the aircraft, not the pilot's head, see page 166).

a **Objectives**: This test checks the pilot's ability to use the display to recover from a rapid hovering turn with sufficient precision to fire a weapon.

b **Requirements**: This evaluation can be flown using any convenient target.

c **Action**: From a stabilized hover at an altitude of less than 20 ft, complete a 180 degree turn. Turns shall be completed in both directions, in a moderate wind from the most critical azimuth. If a critical azimuth has not been defined, the turn shall be completed with the wind blowing directly from the rear of the aircraft. The final heading tolerance should be based on a sight mounted on the aircraft, preferably, the same sight to be used for operational missions.

If a HMD is being evaluated, the turns should be limited to 90 degrees. The pilot's head should be more or less aligned with the aircraft longitudinal axis throughout the maneuver.

d **Data**: Aircraft performance data should be recorded as functions of time:

> Airspeed
> Radar altitude.

Aircraft position should be obtained from LIDAR or other optical tracking system. Differential GPS may be used for aircraft position as well,

Heading angle and tracking accuracy shall be recorded, preferably using visual image recording.

Pilot ratings (DRRs and DFRs) should be hand recorded. Voice recording may be used to supplement the hand recorded data.

e **Performance criteria**:

 i *Desired performance*:

- Maintain longitudinal and lateral position of a selected point on the aircraft within 6 ft of a selected point on the ground.
- Maintain altitude within ±3 ft.
- Stabilize final aircraft heading within a tolerance based on the firing constraints of the weapon system to be deployed on the aircraft.
- Complete the turn so that a firing solution has been achieved within 5 seconds from the initiation of the maneuver. For evaluations using a 90 degree turn, complete the turn so that a firing solution is achieved within 3 seconds from the initiation of the maneuver.

 ii *Adequate performance*:

- Maintain longitudinal and lateral position of a selected point on the aircraft within 12 ft of a selected point on the ground.
- Maintain altitude within ±6 ft.
- Stabilize final aircraft heading within ±3 deg.
- Complete the turn so that a firing solution has been achieved within 10 seconds from the initiation of the maneuver. For evaluations using a 90 degree turn, complete the turn so that a

firing solution is achieved within 6 seconds from the initiation of the maneuver.

f **Safety considerations**: These tests are flown in proximity to the ground and obstructions. Standard precautions for low-level flight should be observed.

Turn-to-Target, Modified

This test is only applicable to head-tracked systems. It repeats the turn-to-target test (see page 164) except that the pilot's head shall be turned first to place his LOS on the target prior to turning the aircraft to point at the target (i. e. turn the pilot's head, then align the aircraft).

a **Objectives**: This test checks the pilot's ability to use the display to recover from a rapid hovering turn coupled with rapid head motion with sufficient precision to fire a weapon.

b **Requirements**: This evaluation can be flown using any convenient target.

c **Action**: From a stabilized hover at an altitude of less than 20 ft, acquire a target 90 degrees from the longitudinal axis of the aircraft. While keeping the HMD sighting reticle aligned with the target, complete a 90 deg turn. Turns shall be completed in both directions, in a moderate wind from the most critical azimuth. If a critical azimuth has not been defined, the turn shall be completed with the wind blowing directly from the rear of the aircraft. The final heading tolerance should be based on a sight mounted on the aircraft, preferably, the same sight to be used for operational missions.

d **Data**: Aircraft performance data should be recorded as functions of time:

Airspeed

Radar altitude.

Aircraft position should be obtained from LIDAR or other optical tracking system. Differential GPS may be used for aircraft position as well.

Heading angle and tracking accuracy shall be recorded, preferably using visual image recording.

Pilot ratings (DRRs and DFRs) should be hand recorded. Voice recording may be used to supplement the hand recorded data.

e **Performance criteria:**
  i *Desired performance*:
  - Maintain longitudinal and lateral position of a selected point on the aircraft within 6 ft of a selected point on the ground.
  - Maintain altitude within ±3 ft.
  - Stabilize final aircraft heading within a tolerance based on the firing constraints of the weapon system to be deployed on the aircraft.
  - Complete the turn so that a firing solution has been achieved within 3 seconds from the initiation of the maneuver.

  ii *Adequate performance*:
  - Maintain longitudinal and lateral position of a selected point on the aircraft within 12 ft of a selected point on the ground.
  - Maintain altitude within ±6 ft.
  - Stabilize final aircraft heading within ±3 deg.
  - Complete the turn so that a firing solution has been achieved within 6 sec from the initiation of the maneuver.

f **Safety considerations:** These tests are flown in proximity to the ground and obstructions. Standard precautions for low-level flight should be observed.

Unusual Attitude Recovery

Unusual attitude (UA) recoveries should be evaluated in simulators and in flight The simulator evaluation should encompass all UAs (including the more extreme attitudes), followed by enough aircraft UAs to ensure a valid statistical sample. The degree of UAs should go beyond 90 deg of bank in all aircraft (at least in simulated flight).

a **Objectives:** There are three objectives for this test. The first objective is to check the adequacy of the attitude display. Next, the utility of the display to allow the pilot to recognize and recovery from unusual attitudes.

  Finally, the evaluation will check for undesired pilot-vehicle-display dynamics during extreme conditions.

b **Requirements:** A fully functioning display and a means to prevent the evaluation pilot from viewing the external visual scene. A predetermined number (eight is typical) of pitch and bank entries should be established covering the range of UAs to be flown. These will vary depending on the type of aircraft.

c **Action**: With the aircraft in straight and level flight, the evaluation display should be blanked. The safety pilot flies the aircraft into a predetermined unusual attitude at which point the display is restored and the flight controls are given back to the evaluation pilot for recovery. Past programs have programmed the UA entries into the autopilot to provide consistent entry conditions.(*111*)

Simulator evaluations should simply reset the simulator to the desired UA entry conditions.

The evaluation pilot recovers to straight and level using standard procedures.

d **Data**: Aircraft performance data should be recorded as functions of time:

> Airspeed
> Altitude
> Pitch and bank angles
> Heading
> Control positions
> Event marker to indicate start of the maneuver.

Pilot ratings (DRRs and DFRs) should be hand recorded. Voice recording may be used to indicate the start of the maneuver and to supplement the hand recorded data.

e **Performance criteria**:

i *Desired performance*:

- Reaction time to initial correct control input < 0.7 sec.*
- No initial control reversals.
- Initial control input in accordance with (IAW) published standards. Published standards include instrument flight manuals (*139-140*) aircraft-specific manuals, or equivalent documents.
- No overshoots on recovery.

ii *Adequate performance*:

- Reaction time to initial correct control input < 1.2 sec.*
- Single initial control reversals.

---

* The reaction time criteria for UA recovery is shorter than that used for failure detection. The evaluation pilot is expecting the UA, while the systems or engine failures are unexpected.

- Initial control input IAW published standards. Published standards include instrument flight manuals (*139-140*) aircraft-specific manuals, or equivalent documents.
- Single overshoot on recovery.

f **Safety considerations**: These tests involve a hooded evaluation pilot. Adequate scan of the external visual scene by the safety pilot is essential. Care must be taken to ensure that the vision restriction for the evaluation pilot does not impair the vision of the safety pilot.

The UAs flown in flight must be within the capability of the aircraft. More extreme attitudes should be performed in the simulator.

Vertical Remask

The vertical remask bob-up/bob-down is a maneuver emulating helicopter combat tactics. It is similar to the maneuver used in ADS-33 (*137*) to check the combination of vertical and lateral handling qualities.

a **Objectives**: This task will check the pilot's ability to use the display to accomplish an aggressive vertical descent close to the ground. It will also check the displays ability to combine vertical and lateral aggressive maneuvers.

b **Requirements**: The test course should include markers to indicate the position during the vertical descent and stabilized hover points. The line between the descent point and the stabilized hover point should be indicated. If the display allows, his evaluation should be flown without using these test course artifacts, to determine if the display provides sufficient cues to the maneuver.

c **Action**: From a stabilized hover at 75 ft, remask vertically to an altitude of 25 ft. Than rapidly displace the aircraft laterally 300 ft and stabilize at a new hover position. Accomplish this maneuver to the left and to the right.

d **Data**: Aircraft performance data should be recorded as functions of time:

> Airspeed
> Altitude
> Heading angle.

Aircraft position should be obtained from LIDAR or other optical tracking system. Differential GPS may be used for aircraft position as well.

Pilot ratings (DRRs and DFRs) should be hand recorded. Voice recording may be used to supplement the hand recorded data.

e **Performance criteria**:

i *Desired performance*:

- During the initial stabilized hover, the vertical descent, and final stabilized hover, maintain longitudinal and lateral position of a selected point on the aircraft within 8 ft of a selected point on the ground.
- Maintain altitude after the remask and during the displacement within ±10 ft.
- Maintain heading within ±10 deg.
- Maintain lateral ground track within ±10 ft.
- Achieve a stabilized hover within 5 seconds after reaching final hover position.
- Achieve the final stabilized hover within 15 seconds of initiating the maneuver.
- Achieve the 25 ft altitude within 6 seconds of initiating the maneuver.

ii *Adequate performance*:

- During the initial stabilized hover, the vertical descent, and final stabilized hover, maintain longitudinal and lateral position of a selected point on the aircraft within 12 ft of a selected point on the ground.
- Maintain altitude after the remask and during the displacement within +10/-15 ft.
- Maintain heading within ±15 deg.
- Maintain lateral ground track within ±15 ft.
- Achieve a stabilized hover within 10 seconds after reaching final hover position.
- Achieve the final stabilized hover within 25 seconds of initiating the maneuver.
- Achieve the 25 ft altitude within 6 seconds of initiating the maneuver.

f **Safety considerations**: These tests are flown in proximity to the ground and obstructions. Standard precautions for low-level flight should be observed. Particular care should be taken during the aggressive descent portion of the test.

Yo-Yo, High

The high yo-yo is a maneuver used in air-to-air combat tactics and is described in ADS-33.(*137*) For HMD testing, the aiming reticle on the HMD should be kept on the other aircraft throughout the test. This task is inappropriate for simulator evaluation because of possible interactions between display dynamics and visual scene delays.

a **Objectives**: The main objective is to check the displays utility in allowing the pilot to use the display to during maneuvering at reduced and elevated load factors. A second objective is to check the short term dynamics of the display/aircraft during aggressive aimpoint tracking including ensuring that there are no undesirable cross-coupling between axes.

b **Requirements**: This evaluation requires a cooperative target aircraft.

c **Action**: The maneuver is initiated from level unaccelerated flight with both aircraft at a constant airspeed equal to the V_H of the test aircraft (airspeed for maximum continuous power). The test aircraft is positioned at least 500 ft in trail behind the target aircraft. The target aircraft then decelerates 20 knots to (V_H -20), causing the test aircraft to close on the target. When the range between the two aircraft decreases to approximately 300 ft, the target aircraft initiates a 60 deg banked turn at constant altitude. The test aircraft delays until the line-of-sight reaches 30 deg off the nose, at which time the pilot initiates a climbing turn toward the target, with a nose-up pitch attitude of 15 to 30 deg. The resulting deceleration causes a decrease in the rate of closure from above. When the closure rate is no longer apparent, and the range to the target is approximately 200 to 500 ft, the test aircraft rapidly lowers the nose to achieve a firing solution within weapon system constraints.

d **Data**: Aircraft performance data should be recorded as functions of time:

       Airspeed
       Altitude
       Pitch and bank angles.

Target acquisition and tracking accuracy shall be recorded, preferably using visual image recording.

Pilot ratings (DRRs and DFRs) should be hand recorded. Voice recording may be used to supplement the hand recorded data.

e **Performance criteria**:
  i *Desired performance*:
  - Maintain aimpoint for 7 sec within firing constrains of the weapon system to be deployed on the aircraft.
  - Acquire the target with no tendency for pitch overshoots.
  - No undesirable cross-coupling.

  ii *Adequate performance*:
  - Maintain aimpoint for 4 sec within firing constrains of the weapon system to be deployed on the aircraft.
  - Acquire the target with no tendency for pitch overshoots.
  - No objectionable cross-coupling.

f **Safety considerations**: These tests are aggressive maneuvers flown in proximity to other aircraft. Suitable precautions for formation flight should be observed, including pre-test briefings.

Yo-Yo, Low

The low yo-yo is a maneuver used in air-to-air combat tactics and is described in ADS-33.(*137*) For HMD testing, the aiming reticle on the HMD should be kept on the other aircraft throughout the test. This task is inappropriate for simulator evaluation because of possible interactions between display dynamics and visual scene delays.

a **Objectives**: The main objective is to check the displays utility in allowing the pilot to use the display to during maneuvering at reduced and elevated load factors. A second objective is to check the short term dynamics of the display/aircraft during aggressive aimpoint tracking including ensuring that there are no undesirable cross-coupling between axes.

b **Requirements**: This evaluation requires a cooperative target aircraft capable of maintaining an airspeed 20 knots greater than V_H for the test aircraft. This aircraft need not be a helicopter.

c **Action**: The maneuver is initiated from level unaccelerated flight with both aircraft at a constant airspeed equal to the V_H of the test aircraft (airspeed for maximum continuous power). The test aircraft is positioned approximately 200 ft in trail behind the target aircraft. The target aircraft then accelerates 20 knots to (V_H +20), resulting in an increase in range between the two aircraft. When the range between the two aircraft increases to approximately 300 ft, the target aircraft initiates a 60 deg banked turn at constant altitude. The test aircraft delays until the line-of-sight reaches 30 deg of the nose, at which time the pilot initiates a diving turn toward the target, with a nose-down pitch attitude of 15 to 30 deg. The resulting acceleration causes the test aircraft to begin to close on the target from above. When a rate of closure is apparent, and the range to the target is within 500 ft, the test aircraft rapidly raises the nose and tracks the target to achieve a firing solution within weapon system constraints.

d **Data**: Aircraft performance data should be recorded as functions of time:

> Airspeed
> Altitude
> Pitch and bank angles.

Target acquisition and tracking accuracy shall be recorded, preferably using visual image recording.

Pilot ratings (DRRs and DFRs) should be hand recorded. Voice recording may be used to supplement the hand recorded data.

e **Performance criteria**:

  i *Desired performance*:
  - Maintain aimpoint for 7 sec within firing constrains of the weapon system to be deployed on the aircraft.
  - Acquire the target with no tendency for pitch overshoots.
  - No undesirable cross-coupling.

  ii *Adequate performance*:
  - Maintain aimpoint for 4 sec within firing constrains of the weapon system to be deployed on the aircraft.
  - Acquire the target with no tendency for pitch overshoots.
  - No objectionable cross-coupling.

f **Safety considerations**: These tests are aggressive maneuvers flown in proximity to other aircraft. Suitable precautions for formation flight should be observed, including pre-test briefings.

Precise Mission Task Elements

The following are precise mission task elements (MTEs) for rotary-wing aircraft. In most cases, they were adapted from other test and evaluation documents and modified to examine display characteristics and display/aircraft interactions, rather than aircraft characteristics alone.

Hover

Hovering is a task that encompasses all helicopters (and VTOL aircraft). The precision hover is a task that is used in handling qualities evaluations.(*137*) For display evaluations, the hover task intentionally requires eyes out in different directions.

a **Objectives**: The objectives are to assess the displays utility in providing information to allow the pilot to transition from translating flight to a stabilized hover with precision and a reasonable amount of aggressiveness, and to use the display to maintain precision position, heading, and altitude.

b **Requirements**: The test course consists of a ground marker showing the target hover point. Seventy-five ft in front of the target hover point is an eight ft tall reference symbol. Markers or cones to the side should indicate the target longitudinal position and the desired and adequate longitudinal tolerances. The 45 deg line of sight from the initial point to the target hover point should be indicated by a line of cones beyond the hover point. This is shown in Figure 11.2.

c **Action**: Initiate the maneuver at a groundspeed between 6 and 10 knots, and an altitude less than 20 ft. The aircraft should be brought to a stabilized hover over the target hover point in a single smooth deceleration. The target hover point shall be oriented approximately 45 deg relative to the heading of the aircraft. The target hover point is a repeatable ground-referenced point from which aircraft deviations are measured. The ground track should be such that the aircraft will arrive over the target hover point (see Figure 11.2). The maneuver is to be performed in calm winds and in winds from the most critical direction. If a critical wind direction has not been established, the maneuver shall be accomplished with the wind blowing directly from the rear of the aircraft.

d **Data**: Aircraft performance data should be recorded as functions of time:

> Airspeed
> Altitude

Pitch and bank angles
Heading.

Sight picture should be recorded using visual image recording.

Aircraft position should be obtained from LIDAR or other optical tracking system. Differential GPS may be used for aircraft position as well.

Pilot ratings (DRRs and DFRs) should be hand recorded. Voice recording may be used to supplement the hand recorded data.

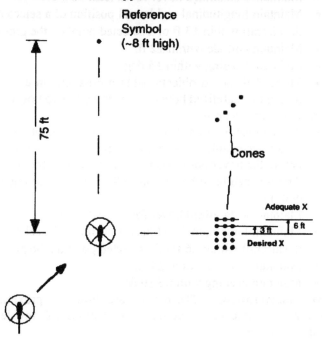

Figure 11.2 Suggested Course for Hover Maneuver
Source: ADS-33 (137)

Note: The hover board to show vertical and lateral position in ADS-33 has been omitted. The display should indicate deviations from the nominal hover point. For non-symbolic displays (such as night vision goggles), the hover board described in ADS-33 (*137*) may be used.

e **Performance criteria**:
  i *Desired performance*:
   - Accomplish the transition to hover in one smooth maneuver. It is not acceptable to accomplish most of the deceleration well before the hover point and then to creep up to the final position.
   - Attain a stabilized hover within 3 sec of the initiation of deceleration.
   - Maintain a stabilized hover for at least 30 sec.
   - Maintain longitudinal and lateral position of a selected point on the aircraft within ±3 ft of a selected point on the ground.
   - Maintain altitude within ±2 ft.
   - Maintain heading within ±5 deg.
   - There shall be no objectionable oscillations in any axis either during the stabilized hover, or the transition to hover.
  ii *Adequate performance*:
   - Accomplish the transition to hover in one smooth maneuver. It is not acceptable to accomplish most of the deceleration well before the hover point and then to creep up to the final position.
   - Attain a stabilized hover within 8 sec of the initiation of deceleration.
   - Maintain a stabilized hover for at least 30 sec.
   - Maintain longitudinal and lateral position of a selected point on the aircraft within ±6 ft of a selected point on the ground.
   - Maintain altitude within ±4 ft.
   - Maintain heading within ±10 deg.
f **Safety considerations**: These tests are flown in proximity to the ground and obstructions. Suitable precautions for low-level flight should be observed.

Hovering Turn

The hovering turn is a rapid turn from a stabilized hover through 180 degrees to another stabilized hover.(*137*)
a **Objectives**: The objectives include checking the display's utility to allowing the pilot to initiate, control, and recover from a moderate rate hovering turn while maintaining position and altitude. An additional objective is to ensure the absence of undesirable cross-coupling.

b **Requirements**: The test course consists of a ground marker showing the target hover point. Markers showing the zero and 180 deg azimuth points should be positioned approximately 100 ft from the target hover point.

c **Action**: From a stabilized hover at an altitude of less than 20 ft, complete a 180 deg turn. Perform the maneuver in both directions, with a moderate wind from the most critical azimuth. If a critical azimuth has not been defined, the turn shall be terminated with the wind blowing directly from the rear of the aircraft.

d **Data**: Aircraft performance data should be recorded as functions of time:

> Airspeed
> Altitude
> Pitch and bank angles
> Heading.

Sight picture should be recorded using visual image recording.

Aircraft position should be obtained from LIDAR or other optical tracking system. Differential GPS may be used for aircraft position as well,

Pilot ratings (DRRs and DFRs) should be hand recorded. Voice recording may be used to supplement the hand recorded data.

e **Performance criteria**:

i *Desired performance*:

- Maintain longitudinal and lateral position of a selected point on the aircraft within ±3 ft of a selected point on the ground.
- Maintain altitude within ±3 ft.
- Stabilize final aircraft heading at 180 deg from the initial heading within ±3 deg.
- Complete the turn to a stabilized hover (within the ±3 deg window) within 10 sec from initiation of the maneuver.

ii *Adequate performance*:

- Maintain longitudinal and lateral position of a selected point on the aircraft within ±6 ft of a selected point on the ground.
- Maintain altitude within ±6 ft.
- Stabilize final aircraft heading at 180 deg from the initial heading within ±6 deg.
- Complete the turn to a stabilized hover (within the ±6 deg window) within 15 sec from initiation of the maneuver.

f **Safety considerations**: These tests are flown in proximity to the ground and obstructions. Suitable precautions for low-level flight should be observed.

Instrument Maneuvers

These maneuvers are standard instrument training maneuvers. These tasks are flown to evaluate the aircraft display symbology and dynamics during tracking. Instrument approaches and terminal area maneuvering will evaluate the cognitive aspects of the display (Chapter 13, page 209).

a **Objectives**: This flight test includes a series of basic maneuvers designed to evaluate the display symbology and to demonstrate that maneuvers can be performed within the performance tolerances appropriate to the maneuvers.

b **Requirements**: These test should be flown in appropriate aircraft configurations. These tests should normally be flown in actual or simulated IMC.

c **Action**: The representative instrument tasks from the following list should be flown using normal procedures from instrument training manuals (*139-140*).

  i *vertical S-A*
  ii *vertical S-C*
  iii *vertical S-D*
  iv *oboe patterns*
  v *yankee patterns*
  vi *steep turns* (bank depending on aircraft type, through 360 deg, immediate reversal 360 deg back to original heading).

d **Data**: Aircraft performance data should be recorded as functions of time:

        Airspeed
        Barometric altitude
        Radar altitude
        Heading
        Pitch and bank angles.

Pilot ratings (DRRs and DFRs) should be hand recorded. Voice recording may be used to supplement the hand recorded data.

e **Performance criteria**:
  i *Desired performance*:
   - Maintain altitude within ±100 ft during level legs; ±200 ft during level-offs.
   - Maintain heading ±3 deg during straight legs; ±5 deg rolling out.
   - Maintain airspeed ±5 knots.
   - Complete all legs within ±5 sec of standard times.
   - Maintain bank (steep turns) within ±3 deg.
   - Maintain airspeed (steep turns) ±5 knots.

  ii *Adequate performance*:
   - Maintain altitude within ±200 ft during level legs; ±300 ft during level-offs.
   - Maintain heading ±5 deg during straight legs; ±10 deg rolling out.
   - Maintain airspeed ±5 knots.
   - Complete all legs within ±10 sec of standard times.
   - Maintain bank (steep turns) within ±5 deg.
   - Maintain airspeed (steep turns) ±10 knots.

f **Safety considerations**: These tests involve a hooded evaluation pilot. Adequate scan of the external visual scene by the safety pilot is essential. Care must be taken to ensure that the vision restriction for the evaluation pilot does not impair the vision of the safety pilot.

Instrument Tracking

The evaluations under this MTE are designed solely to evaluate the ability of the pilot to use the display under tracking conditions. Instrument approaches will be evaluated more completely under mission related tasks that will include evaluation of situation awareness, cognitive processing, and other mission tasks.

a **Objectives**: This flight test includes a series of basic maneuvers designed to evaluate the display symbology and to check for pilot-aircraft-display dynamic issues arising during tight compensatory tracking tasks.

b **Requirements**: These test should be flown in appropriate aircraft configurations. These tests should normally be flown in actual or simulated IMC.

c **Action**: An appropriate radio navigation station (VOR, TACAN, or ADF, as appropriate for the installed equipment) should be tuned. From a 45 deg intercept leg, intercept the selected courses. Track both inbound and outbound while maintaining altitude. The tracking should include some station passages.

   Tune an ILS facility (if an appropriate receiver is installed). From a 45 deg intercept leg, intercept the selected course. From inbound tracking in the ILS localizer front course, intercept the glideslope and track down toward the runway. Backcourse localizer courses should also be flown.

d **Data**: Aircraft performance data should be recorded as functions of time:

>   Airspeed
>   Barometric altitude
>   Radar altitude
>   Heading
>   Course deviation
>   Glideslope deviation
>   Marker beacon signal.

   Pilot ratings (DRRs and DFRs) should be hand recorded. Voice recording may be used to supplement the hand recorded data.

e **Performance criteria**:

   i *Desired performance*:
   - Maintain altitude within ±100 ft during level legs.
   - Maintain heading (during intercepts) ±3 deg.
   - Maintain airspeed ±5 knots.
   - Tracking within ±1/2 dot for fifty percent of tracking task.
   - Tracking within ±1 dot for entire of tracking task.
   - No overshoots (beyond 1/2 dot) on course intercept.

   ii *Adequate performance*:
   - Maintain altitude within ±200 ft during level legs.
   - Maintain heading (during intercepts) ±5 deg.
   - Maintain airspeed ±5 knots.
   - Tracking within ±1 dot for fifty percent of tracking task.
   - Tracking within ±2 dot for entire of tracking task.
   - Single overshoot on course intercept.

f **Safety considerations**: These tests involve a hooded evaluation pilot. Adequate scan of the external visual scene by the safety pilot is essential. Care must be taken to ensure that the vision restriction for the evaluation pilot does not impair the vision of the safety pilot.

Landing

The ADS-33 landing maneuver is an evaluation of the pilot-aircraft-display in a smooth transition from flight to landing.(*137*) It is an excellent maneuver for evaluating a helicopter for confined-space operations.

a **Objectives**: There are two objectives. The first is to evaluate the pilot's ability to use the display to control aircraft position during the final descent to a precision landing point. A second objective checks pilot-vehicle-display dynamics when the pilot is forced into tight compensatory tracking behavior.

b **Requirements**: The test course consists of a raised rectangular landing platform that exceeds the dimensions of the aircraft landing gear by 3 ft in width and 6 ft in length. The platform shall be raised above the surrounding surface by an amount that will result in an obvious deck angle if the aircraft is landed with any part of the landing gear off the platform. However, the height of the platform should be less than a value that would cause the aircraft to exceed its slope landing limits if it is landed with any part of the landing gear off the platform.

c **Action**: Starting from an altitude of greater than 10 ft, maintain an essentially steady descent to the landing platform. It is acceptable to arrest sink rate momentarily to make last minute corrections before touchdown.

d **Data**: Aircraft performance data should be recorded as functions of time:

>   Airspeed
>   Altitude
>   Pitch and bank angles
>   Heading.

Final landing position should be obtained by direct measurement. Aircraft position should be obtained from LIDAR or other optical tracking system. Differential GPS may be used for aircraft position as well,

Pilot ratings (DRRs and DFRs) should be hand recorded. Voice recording may be used to supplement the hand recorded data.

e **Performance criteria**:
  i *Desired performance*:
  - Accomplish the landing with a smooth continuous descent with no objectionable oscillations.
  - Once the altitude is below 10 ft, complete the landing within 10 sec.
  - Touchdown within ±0.5 ft laterally and ±1 ft longitudinally of a designated position on the platform.
  - Attain a aircraft heading at touchdown that is aligned with the longitudinal axis of the platform within ±5 deg.
  - The final position of the aircraft shall be the position that existed at touchdown. It is not acceptable to adjust the aircraft position and heading after all elements of the landing gear have made contact with the platform.
  ii *Adequate performance*:
  - Touchdown on and remain on the platform.
  - Attain a aircraft heading at touchdown that is aligned with the longitudinal axis of the platform within ±10 deg.
f **Safety considerations**: These tests are flown in proximity to the ground and obstructions. Suitable precautions for low-level flight should be observed. Particular care must be taken to ensure that landing with part of the landing gear off and part on the platform does not constitute an unreasonable hazard.

Pirouette

The pirouette is a maneuver used in helicopter handling qualities evaluation.(*137*) It is an excellent maneuver for evaluating a helicopter for confined-space operations. Because it requires the pilot to shift his gaze back and forth from straight ahead to ninety-degrees off-axis, it will identify problems in the head-tracker system.

The pirouette is not suitable for head-down display evaluation.

a **Objectives**: The pirouette is intended to check the pilot's ability to use the display to accomplish precision control of the aircraft simultaneously in pitch, roll, yaw, and heave.
b **Requirements**: The test course consists of markings on the ground that clearly denote a 100 ft radius circular pathway. For non-symbology displays (such as NVGs), reference markings to indicate acceptable and

adequate performance boundaries should be added. The suggested course is shown in Figure 11.3.

c **Action**: Initiate the maneuver from a stabilized hover over a point on the circumference of a 100 ft radius circle with the nose of the aircraft pointed at the reference point at the center of the circle, and a hover altitude of 10 ft. Accomplish a lateral translation around the circle, keeping the nose of the aircraft pointed at the center of the circle, and the circumference of the circle under a selected point on the aircraft. Terminate the maneuver over the starting point. Perform the maneuver in both directions. Perform the maneuver in calm or light winds.

d **Data**: Aircraft performance data should be recorded as functions of time:

> Airspeed
> Altitude.

Tracking accuracy should be recorded using visual image recording.

Aircraft position should be obtained from LIDAR or other optical tracking system. Differential GPS may be used for aircraft position as well,

Pilot ratings (DRRs and DFRs) should be hand recorded. Voice recording may be used to supplement the hand recorded data.

e **Performance criteria**:

i *Desired performance*:

- Maintain a selected reference point on the aircraft within ±10 ft of the circumference of the circle.
- Maintain altitude within ±3 ft.
- Maintain heading so that the nose of the aircraft points at the center of the circle within ±10 deg.
- Complete the circle and arrive back over the starting point within 45 sec. Maintain essentially constant lateral groundspeed throughout the maneuver. (The nominal lateral velocity will be approximately 8 knots.)
- Maintain a stabilized hover (within 10 ft) within 5 sec after returning to the starting point.

ii *Adequate performance*:

- Maintain a selected reference point on the aircraft within ±15 ft of the circumference of the circle.
- Maintain altitude within ±10 ft.

- Maintain heading so that the nose of the aircraft points at the center of the circle within ±15 deg.
- Complete the circle and arrive back over the starting point within 60 sec. Maintain essentially constant lateral ground-speed throughout the maneuver. (The nominal lateral velocity will be approximately 6 knots.)
- Maintain a stabilized hover (within 15 ft) within 10 sec after returning to the starting point.

f **Safety considerations:** These tests are flown in proximity to the ground and obstructions. Suitable precautions for low-level flight should be observed.

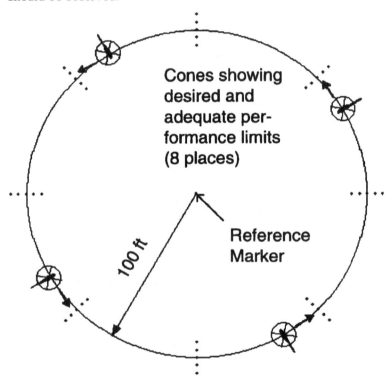

Cones showing desired and adequate performance limits (8 places)

100 ft

Reference Marker

Figure 11.3 Pirouette Course
Source: *ADS-33 (137)*

12 Fixed-Wing Mission Task Elements

Aggressive Mission Task Elements

The following are aggressive mission task elements (MTEs) for fixed-wing aircraft. In most cases, they were adapted from other test and evaluation documents and modified to examine display characteristics and display/aircraft interactions, rather than aircraft characteristics alone. In some cases, the flight in the clear-visual-day environment will seem trivial; however, such evaluations will form the basis for subsequent night or adverse weather evaluations.

Acrobatics

This flight test includes a series of basic maneuvers designed to evaluate the display symbology and to demonstrate that maneuvers can be performed within the performance tolerances appropriate to the maneuvers.

a **Objectives**: The primary objective is to demonstrate the absence of undesirable interactions and cross-couplings during dynamic maneuvering. Secondarily, the objective is to demonstrate the utility of the display in performing these maneuvers and to demonstrate that maneuvers can be performed within the performance tolerances appropriate to the maneuvers.

b **Requirements**: There no particular test requirements for these evaluations.

c **Action**: The representative flight maneuvers from the following list should be flown using normal procedures.
  i *Barrel roll.*
  ii *Cloverleaf.*
  iii *Modified lazy-eight* Perform rolling pull-up so that the nose passes a heading 30 deg from the initial heading at 30 deg nose-high. Continue the turn passing the horizon wings-level inverted 60 deg from

185

the entry heading. Reverse the direction of the roll so that the nose passes the 90 deg heading change at 30 deg nose down. Continue the maneuver with the nose reaching the horizon, wings-level 120 deg from the initial heading. Continue the roll and pull to reverse the maneuver back to the original heading. The altitudes as the nose passes the horizon should be the same as the entry altitude.

    iv *Steep turns* (constant bank turns through 360 deg, immediate reversal 360 deg back to original heading. The bank angle should be appropriate for the aircraft type).

d **Data**: Aircraft performance data should be recorded as functions of time:

> Airspeed
> Altitude
> Pitch and bank angles
> Heading.

Pilot ratings (DRRs and DFRs) should be hand recorded. Voice recording may be used to supplement the hand recorded data.

e **Performance criteria**:

    i *Desired performance*:
- Achieve key points within ±200 ft altitude.
- Achieve key points within ±10 deg heading.
- Achieve key points within ±10 deg bank.
- Achieve key points within ±10 knots airspeed.
- Maintain bank (steep turns) within ±3 deg.
- Maintain airspeed (steep turns) ±5 knots.

    ii *Adequate performance*:
- Achieve key points within ±400 ft altitude.
- Achieve key points within ±20 deg heading.
- Achieve key points within ±20 deg bank.
- Achieve key points within ±20 knots airspeed.
- Maintain bank (steep turns) within ±10 deg.
- Maintain airspeed (steep turns) ±10 knots.

f **Safety considerations**: There are no particular hazards associated with these tests. The aircraft must be flown within its operational limits. The pilots must maintain an adequate watch for other air traffic.

Aggressive Air-to-Ground Visual Tracking

Air-to-ground visual tracking as an extremely useful test to determine if there are any problems with the short period dynamics of a see-through display during aggressive maneuvering.(*136*) This task should be performed on all aircraft equipped with HUDs or HMDs regardless of their intended missions. Aggressive visual tracking is inappropriate for simulator evaluation because of possible interactions between display dynamics and visual scene delays.

a **Objectives**: The objective is to check pilot-vehicle-display dynamics when the pilot is forced into tight compensatory tracking behavior.

b **Requirements**: This evaluation requires a convenient ground target.

c **Action**: The maneuver is initiated from descending flight toward the target at the desired airspeed. For all aircraft, landing reference speed, approximately a 3 deg descent, and a three mile approach leg should be used. For air-to-ground aircraft, weapon delivery airspeed, a suitable dive angle, and a suitable distance should also be used. The test aircraft is flown toward the target maintaining the aiming symbol (or aircraft reference symbol) on the ground target. At the approximate halfway point, the aimpoint should be shifted to a new aimpoint. The maneuver should be discontinued at a safe altitude.

d **Data**: Aircraft performance data should be recorded as functions of time:

> Airspeed
> Altitude
> Pitch and bank angles.

Target acquisition and tracking accuracy shall be recorded, preferably using visual image recording.

Pilot ratings (DRRs and DFRs) should be hand recorded. Voice recording may be used to supplement the hand recorded data.

e **Performance criteria**:

i *Desired performance*:

- Acquire new target with no tendency for overshoots.
- Maintain aimpoint within 5 mrad or within firing constrains of the weapon system to be deployed on the aircraft.
- Acquire new target with no tendency for overshoots.
- Maintain aimpoint within 5 mrad or within firing constrains of the weapon system to be deployed on the aircraft.

- No undesirable cross-coupling.
  ii *Adequate performance*:
   - Acquire new target with minimal tendency for overshoots.
   - Maintain aimpoint within 10 mrad or within firing constrains of the weapon system to be deployed on the aircraft.
   - Acquire new target with minimal tendency for overshoots.
   - Maintain aimpoint within 10 mrad or within firing constrains of the weapon system to be deployed on the aircraft.
   - No objectionable cross-coupling.
f **Safety considerations**: These tests are aggressive maneuvers flown in proximity to the ground and obstructions. Suitable precautions for low-level flight should be observed.

Air-to-Air Visual Tracking

This task is representative of air combat maneuvering (ACM). It is inappropriate for simulator evaluation because of possible interactions between display dynamics and visual scene delays.

a **Objectives**: The objective is to evaluate the display's short term dynamics during aggressive aimpoint tracking. It is important to remember that the evaluation is of the display, not the pilot. For this reason, all maneuvers are in level flight. The evaluation pilot will not know the time of the maneuver, although he will know the direction and magnitude.

b **Requirements**: This evaluation requires a cooperative target aircraft of comparable performance.

c **Action**: The maneuver is initiated from level unaccelerated flight with both aircraft at the corner speed of the test aircraft. The test aircraft is positioned 500 to 1000 ft in trail behind the target aircraft. The target aircraft then performs a wind-up during increasing to 5 g (or other suitable value). The evaluation pilot tracks the lead aircraft throughout. When ten seconds of tracking at the final turn rate is achieved, the evaluation pilot will make a radio call to that effect. After a short delay, the lead aircraft reverse the turn to the same bank angle in the opposite direction. The test aircraft follows, maintaining the aiming sight on the target aircraft.

Repeat the maneuver in the opposite direction.

Several replications of these flight envelope limit data points will be required.

d **Data**: Aircraft performance data should be recorded as functions of time:

> Airspeed
> Altitude
> Pitch and bank angles.

Target acquisition and tracking accuracy shall be recorded, preferably using visual image recording.

Pilot ratings (DRRs and DFRs) should be hand recorded. Voice recording may be used to supplement the hand recorded data.

e **Performance criteria**:

i *Desired performance*:

- Maintain aimpoint for seventy-five percent of time prior to turn reversal within firing constrains of the weapon system to be deployed on the aircraft.
- Acquire the target after the reversal within 8 sec with no tendency for pitch overshoots.
- No undesirable cross-coupling.

ii *Adequate performance*:

- Maintain aimpoint for fifty percent of time prior to turn reversal within firing constrains of the weapon system to be deployed on the aircraft.
- Acquire the target after the reversal within 12 sec with no tendency for pitch overshoots.
- No undesirable cross-coupling.

f **Safety considerations**: These tests are aggressive maneuvers flown in proximity to other aircraft. Suitable precautions for formation flight should be observed, including pre-test briefings.

Engine Failure

These test are intended for multiengine airplanes, particularly those with wing-mounted engines. These tests shall be flown in conjunction with other evaluations being conducted in order to present the evaluation pilot with unexpected failures.

a **Objectives**: These tests are intended to verify that the display will assist the pilot in recognizing engine failures and taking appropriate corrective action while maintaining aircraft control.

b **Requirements**: Engine failures should be introduced during mission related tasks to determine the reaction time and responses of the pilot. Appropriate procedures for operating the aircraft following these failures must be established prior to conducting these tests.

c **Action**: For tests involving operations in IMC, the evaluation pilot shall be hooded or otherwise prevented from using outside visual cues. Engine failures shall be introduced during operations at unexpected times by the safety pilot.

Upon recognition of the failure, the evaluation pilot shall take appropriate action as described in the aircraft/system operating procedures. The evaluation pilot will be briefed to make a verbal call upon recognition of all system failures as a means of determining detection time.

Following recovery, the evaluation pilot shall complete displays ratings (DRR and DFR).

d **Data**: Aircraft performance data should be recorded as functions of time:

> Airspeed
> Altitude
> Pitch and bank angles
> Control positions
> Event (for start of maneuver).

Pilot ratings (DRRs and DFRs) should be hand recorded. Voice recording may be used to indicate the start of the UA and to used to supplement the hand recorded data.

e **Performance criteria**:

i *Desired performance*:

- Initial correct engine control input within 4 sec.
- Initial engine control input in accordance with published standards. Published standards include aircraft-specific manuals or equivalent documents.
- Maintain altitude within ±200 ft.
- Maintain airspeed within ±5 knots.
- Maintain heading within ±10 deg.

ii *Adequate performance*:
  - Initial correct engine control input within 8 sec.
  - Initial engine control input in accordance with published standards. Published standards include aircraft-specific manuals or equivalent documents.
  - Maintain altitude within ±400 ft.
  - Maintain airspeed within ±10 knots.
  - Maintain heading within ±30 deg.

f **Safety considerations**: These tests involve simulated engine failure and shut-down. Suitable precautions for such maneuvers should be observed.

Flight Control System Failures

These tests are intended to evaluate the ability of the pilot to use the display to recover from flight control failures. Situation awareness and workload will be evaluated in mission-related tests. These types of failures are adapted from autopilot failure testing.(*63*) The general method is to introduce the failure at unexpected times and determine the resultant deviation from the desired aircraft flight path.

a **Objectives**: These tests are designed to verify that the pilot can recognize critical flight control system (FCS), stability augmentation system (SAS), autopilot system, and display dynamic system failures and take appropriate corrective action while maintaining aircraft control.

b **Requirements**: These tests shall be flown in conjunction with other evaluations being conducted.

Engine failures should be introduced during mission related tasks to determine the reaction time and responses of the pilot. Appropriate procedures for operating the aircraft following these failures must be established prior to conducting these tests.

FCS, SAS, autopilot, and display system failures shall be introduced during mission related tasks to determine the reaction time and responses of the pilot. Appropriate procedures for operating the aircraft following these failures shall be established prior to conducting these tests.

It may be required to use a special software version to allow the flight test engineer to introduce some system failures such as simulating attitude gyro failures. If such software versions are used, the response of the system to the simulated failure shall be verified prior to conducting these tests.

Typical failures include

- FCS failures (change in gains or loss of function)
- SAS failures (hardover or loss of SAS function)
- Autopilot failure (both hardover and slow failures)
- Loss of quickening or change in quickening function
- Change in display damping.

c **Action**: For tests involving operations in IMC, the evaluation pilot shall be hooded or otherwise prevented from using outside visual cues. System failures shall be introduced during operations at unexpected times by the safety pilot or flight test engineer.

Upon recognition of the failure, the evaluation pilot shall take appropriate action as described in the aircraft/system operating procedures. The evaluation pilot will be briefed to make a verbal call upon recognition of all system failures as a means of determining detection time.

Autopilot testing normally requires a delay (one to three seconds) after recognition before the pilot takes corrective action. These delays depend on the degree of pilot attention. In the systems failure tests covered in this section, no such recognition time will be required since the pilot will be actively involved in flight control.

Following recovery, the evaluation pilot shall complete displays' ratings (DRR and DFR).

d **Data**: Aircraft performance data should be recorded as functions of time:

Airspeed
Altitude
Pitch and bank angles
Control positions
Parameters appropriate to failure
Event (for start of maneuver).

Pilot ratings (DRRs and DFRs) should be hand recorded. Voice recording may be used to indicate the start of the UA and to used to supplement the hand recorded data.

e **Performance criteria**:
  i *Desired performance*:
  - Failure detection time within 2 sec of first indication. Slowly developing failures, such as autopilot softovers may require addition time. In such cases, it may be better to describe the desired and adequate performance in terms of flight path deviations rather than time.
  - Pilot actions in accordance with published standards. Published standards include aircraft-specific manuals or equivalent documents.
  - Maintain altitude within ±200 ft.
  - Maintain airspeed within ±10 knots.
  - Maintain heading within ±10 deg.
  ii *Adequate performance*:
  - Failure detection time within 5 sec of first indication. Slowly developing failures, such as autopilot softovers may require addition time. In such cases, it may be better to describe the desired and adequate performance in terms of flight path deviations rather than time.
  - Pilot actions in accordance with published standards. Published standards include aircraft-specific manuals or equivalent documents.
  - Maintain altitude within ±500 ft.
  - Maintain airspeed within ±20 knots.
  - Maintain heading within ±20 deg.
f **Safety considerations**: These tests involve simulated flight control failures. Suitable precautions for should be observed.

Formation

This test is an evaluation of the ability of the display to aid formation flight in adverse weather. In good visual conditions, it is simply an evaluation of non-interference with formation flight. This task is inappropriate for simulator evaluation because of possible interactions between display dynamics and visual scene delays.

These tasks are intended for evaluation in fighter aircraft, not in large formations of transport or bomber airplanes. Formation for those aircraft are considered to be more cognitive in nature rather than dynamic.

Because of this, they will be flown under mission tasks in the following chapter (Chapter 13).

Refueling (in a receiving aircraft) is covered separately.

a **Objectives**: There are several objectives for this task. First, to ensure non-interference by the display during formation flight in clear conditions. Second, to evaluate the display's utility during formation maneuvers in degraded weather.

Remaining objectives include evaluating the short term dynamics of the display/aircraft during aggressive maneuvers. and checking for undesirable cross-coupling between axes.

b **Requirements**: This evaluation requires a cooperative aircraft of comparable performance. Both pilots should be briefed on the flight envelope: minimum altitudes, maximum bank angles, and minimum and maximum airspeeds.

c **Action**: The maneuver is initiated from level unaccelerated flight with both aircraft at the corner speed of the test aircraft. The test aircraft is positioned on the wing of the lead aircraft. The lead aircraft then performs a series of level, descending, and climbing turns while varying airspeed. The test aircraft follows, maintaining its position on the lead aircraft. During the evaluation, the test aircraft should cross-under to the opposite wing.

During the exercise, the lead aircraft should approach the limits of the flight envelope without penetrating the limits. For some runs, the lead aircraft will fly through these limits. In such cases, the evaluation pilot will call "knock it off" and recover to wings level flight.

d **Data**: Aircraft performance data should be recorded as functions of time:

> Airspeed
> Altitude
> Pitch and bank angles
> Control positions.

Pilot ratings (DRRs and DFRs) should be hand recorded. Voice recording will record recognition of achieving flight envelope limits ("knock it off") and may be used to supplement the hand recorded data.

e **Performance criteria**:

i *Desired performance*:

- Maintain position within satisfactory tolerances for aircraft type.

- Maintain satisfactory altitude awareness
  (abandon task at minimum altitude +/-200 ft).
- Maintain satisfactory attitude awareness
  (abandon task at maximum bank angle +/-5 deg).
- Maintain satisfactory airspeed awareness
  (abandon task at airspeed limits +/-10 knots).
- No undesirable interference with view of lead aircraft.
- No undesirable pilot-vehicle-display dynamics.

ii *Adequate performance*:
- Maintain position within satisfactory tolerances for aircraft type.
- Maintain satisfactory altitude awareness
  (abandon task at minimum altitude +/-200 ft).
- Maintain satisfactory attitude awareness
  (abandon task at maximum bank angle +/-5 deg).
- Maintain satisfactory airspeed awareness
  (abandon task at airspeed limits +/-10 knots).
- No objectionable interference with view of lead aircraft.
- No objectionable pilot-vehicle-display dynamics.

f **Safety considerations**: These tests are aggressive maneuvers flown in proximity to other aircraft. Suitable precautions for formation flight should be observed, including pre-test briefings. It may be desirable to fly initial trials in trail formation. Caution must be exercised in approaching or exceeding maximum bank angles while turning into the wingman (test aircraft).

Refueling

This test is an evaluation of the ability of the display to aid the pilot of the receiving aircraft during refueling in adverse weather. In good visual conditions, it is simply an evaluation of non-interference with refueling operations. This task is inappropriate for simulator evaluation because of possible interactions between display dynamics and visual scene delays.

These tasks need only be flown in aircraft which intended to receive fuel from tankers.

a **Objectives**: There are several objectives for this task. First, to ensure non-interference by the display during refueling flight in clear conditions. Second, to evaluate the display's utility during refueling maneuvers in degraded weather.

Remaining objectives include evaluating the short term dynamics of the display/aircraft and checking for undesirable cross-coupling between axes.

b **Requirements**: This evaluation requires a cooperative tanker. Both pilots should be briefed on the flight envelope: minimum altitudes, maximum bank angles, and minimum and maximum airspeeds.

It is not necessary to transfer fuel; dry contacts are acceptable.

c **Action**: The maneuver is initiated with the test aircraft at the precontact position. Both aircraft should be stabilized at the appropriate airspeed. The tanker should perform normal tanker tracks including level turns to remain in a refueling area. Both left and right turns should be flown.

For boom-and-receptacle operations, the receiver should maneuver within the boom envelope. Where appropriate, descending refueling operations ("Toboggan") should be flown.

At least one simulated breakaway should be flown.

d **Data**: Aircraft performance data should be recorded as functions of time:

> Airspeed
> Altitude
> Pitch and bank angles
> Control positions
> Boom angle and extension (boom tanker only).

Pilot ratings (DRRs and DFRs) should be hand recorded. Voice recording may be used to supplement the hand recorded data.

e **Performance criteria**:

i *Desired performance*:

- Maintain position within satisfactory tolerances for aircraft type.
- No undesirable interference with view of lead aircraft.
- No undesirable pilot-vehicle-display dynamics.

ii *Adequate performance*:

- Maintain position within satisfactory tolerances for aircraft type.
- No objectionable interference with view of lead aircraft.
- No objectionable pilot-vehicle-display dynamics.

f **Safety considerations**: These tests are aggressive maneuvers flown in proximity to other aircraft. Suitable precautions for formation flight should be observed, including pre-test briefings.

Steep Turns and Reversals

These tasks are intended for non-acrobatic aircraft, such as transports, tankers, or bombers.

a **Objectives**: This flight test includes a series of basic maneuvers designed to evaluate the display symbology and to demonstrate that maneuvers can be performed within the performance tolerances appropriate to the maneuvers.

b **Requirements**: There no particular test requirements for these evaluations.

c **Action**: Steep turns through 360 degrees of heading should be flown followed by immediate reversal through 360 degrees back to the original heading. The bank angle should be appropriate to the aircraft type.

d **Data**: Aircraft performance data should be recorded as functions of time:

> Airspeed
> Altitude
> Pitch and bank angles
> Heading.

Pilot ratings (DRRs and DFRs) should be hand recorded. Voice recording may be used to supplement the hand recorded data.

e **Performance criteria**:

i *Desired performance*:
- Maintain altitude within ±200 ft.
- Roll out within ±10 deg of target heading.
- Maintain bank (steep turns) within ±3 deg.
- Maintain airspeed (steep turns) ±5 knots.

ii *Adequate performance*:
- Maintain altitude within ±400 ft.
- Roll out within ±20 deg of target heading.
- Maintain bank (steep turns) within ±10 deg.
- Maintain airspeed (steep turns) ±10 knots.

f **Safety considerations**: There are no particular hazards associated with these tests. The aircraft must be flown within its operational limits. The pilots must maintain an adequate watch for other air traffic.

Turn-to-Target

This task is intended for aircraft with an air-to-ground weapon delivery capability. For helmet-mounted display evaluations, the pilot's line-of-sight should not remain fixed on the target during the turn-to-target. Because of difficulties with simulator dynamics and visual scene synchronization. this task is not suitable for simulation.

a **Objectives**: Check pilot's ability to use the display to rapidly acquire a target with sufficient precision to fire a weapon.

b **Requirements**: This evaluation requires an air-to-ground weapons range.

c **Action**: From straight and level flight with the target 6000 to 15000 ft off the wingtip, aggressively turn the aircraft toward the target and achieve a targeting solution. Track the target for at least seven seconds. The initial altitude should be appropriate to the intended weapon delivery dive angle. Turns shall be completed in both directions.

If a HMD is being evaluated, the pilot's head should be more or less aligned with the aircraft longitudinal axis throughout the maneuver.

d **Data**: Aircraft performance data should be recorded as functions of time:

> Airspeed
> Altitude
> Pitch and bank angles
> Control positions.

Target acquisition and tracking accuracy shall be recorded, preferably using visual image recording.

Pilot ratings (DRRs and DFRs) should be hand recorded. Voice recording may be used to supplement the hand recorded data.

e **Performance criteria**:

i *Desired performance*:

- Stabilize final aircraft heading within a tolerance based on the firing constraints of the weapon system to be deployed on the aircraft.
- Maintain airspeed within ±10 knots.

- Complete the turn so that a firing solution has been achieved within 5 seconds from the initiation of the maneuver.

ii *Adequate performance*:

- Stabilize final aircraft heading within a tolerance based on the firing constraints of the weapon system to be deployed on the aircraft.
- Maintain airspeed within ±20 knots.
- Complete the turn so that a firing solution has been achieved within 10 seconds from the initiation of the maneuver.

f **Safety considerations**: These tests are flown in proximity to the ground and obstructions. Standard precautions for low-level flight should be observed.

Turn-to-Target, Modified

This task is intended for aircraft with an air-to-ground weapon delivery capability. For helmet-mounted display evaluations, the pilot's line-of-sight should remain fixed on the target during the turn-to-target. Because of difficulties with simulator dynamics and visual scene synchronization, this task in not suitable for simulation.

a **Objectives**: Check pilot's ability to use the display to rapidly acquire a target with sufficient precision to fire a weapon.

b **Requirements**: This evaluation requires an air-to-ground weapons range.

c **Action**: From straight and level flight with the target 6000 to 15000 ft off the wingtip, turn the pilot's line-of-sight (LOS) to place the aiming reticle on the target, then aggressively turn the aircraft toward the target and achieve a targeting solution. Track the target for at least seven seconds. The initial altitude should be appropriate to the intended weapon delivery dive angle. Turns shall be completed in both directions.

The pilot's LOS should be kept on the target throughout the maneuver.

d **Data**: Aircraft performance data should be recorded as functions of time:

> Airspeed
> Altitude
> Pitch and bank angles
> Control positions.

Target acquisition and tracking accuracy shall be recorded, preferably using visual image recording.

Pilot ratings (DRRs and DFRs) should be hand recorded. Voice recording may be used to supplement the hand recorded data.

e **Performance criteria**:
   i *Desired performance*:
   - Stabilize final aircraft heading within a tolerance based on the firing constraints of the weapon system to be deployed on the aircraft.
   - Maintain airspeed within ±10 knots.
   - Complete the turn so that a firing solution has been achieved within 5 seconds from the initiation of the maneuver.
   ii *Adequate performance*:
   - Stabilize final aircraft heading within a tolerance based on the firing constraints of the weapon system to be deployed on the aircraft.
   - Maintain airspeed within ±20 knots.
   - Complete the turn so that a firing solution has been achieved within 10 seconds from the initiation of the maneuver.

f **Safety considerations**: These tests are flown in proximity to the ground and obstructions. Standard precautions for low-level flight should be observed.

Unusual Attitude Recovery

Unusual attitude (UA) recoveries should be evaluated in simulators and in flight The simulator evaluation should encompass all UAs (including the more extreme attitudes), followed by enough aircraft UAs to ensure a valid statistical sample. The degree of UAs should go beyond 90 deg of bank in all aircraft (at least in simulated flight).

a **Objectives**: There are three objectives for this test. The first objective is to check the adequacy of the attitude display. Next, the utility of the display to allow the pilot to recognize and recovery from unusual attitudes.

Finally, the evaluation will check for undesired pilot-vehicle-display dynamics during extreme conditions.

b **Requirements**: A fully functioning display and a means to prevent the evaluation pilot from viewing the external visual scene. A predetermined number (eight is typical) of pitch and bank entries should be established covering the range of UAs to be flown. These will vary depending on the type of aircraft.

c **Action**: With the aircraft in straight and level flight, the evaluation display should be blanked. The safety pilot flies the aircraft into a predetermined unusual attitude at which point the display is restored and the flight controls are given back to the evaluation pilot for recovery. Past programs have programmed the UA entries into the autopilot to provide consistent entry conditions.(*111*)

Simulator evaluations should simply reset the simulator to the desired UA entry conditions. An alternative simulator entry has the pilot to fly in trail to a lead aircraft which then performs a series of maneuvers covering the desired envelope of attitudes. At an unexpected time, the lead aircraft is removed from the simulator scene and the pilot recovers to straight-and-level. This approach was used by Gallimore *et al. (114)* Such an UA initiation may be more realistic than the conventional reset initialization.

The evaluation pilot recovers to straight and level using standard procedures.

d **Data**: Aircraft performance data should be recorded as functions of time:

> Airspeed
> Altitude
> Pitch and bank angles
> Control positions
> Event (for start of maneuver).

Pilot ratings (DRRs and DFRs) should be hand recorded. Voice recording may be used to indicate the start of the UA and to used to supplement the hand recorded data.

e **Performance criteria**:

i *Desired performance*:

- Reaction time to initial correct control input < 0.7 sec.*
- No initial control reversals.

---

* The reaction time criteria for UA recovery is shorter than that used for failure detection. The evaluation pilot is expecting the UA, while the systems or engine failures are unexpected.

- Initial control input in accordance with (IAW) published standards. Published standards include instrument flight manuals (*139-140*) aircraft-specific manuals, or equivalent documents.
- No overshoots on recovery.

ii *Adequate performance*:
- Reaction time to initial correct control input < 1.2 sec.
- Single initial control reversals.
- Initial control input IAW published standards. Published standards include instrument flight manuals (*139-140*) aircraft-specific manuals, or equivalent documents.
- Single overshoot on recovery.

f **Safety considerations**: These tests involve a hooded evaluation pilot. Adequate scan of the external visual scene by the safety pilot is essential. Care must be taken to ensure that the vision restriction for the evaluation pilot does not impair the vision of the safety pilot.

The UAs flown in flight must be within the safe capability of the aircraft. More extreme attitudes should be performed in the simulator.

Yo-Yo, High

The high yo-yo is a maneuver used in air-to-air combat tactics. For HMD testing, the aiming reticle on the HMD should be kept on the other aircraft throughout the test. This task is inappropriate for simulator evaluation because of possible interactions between display dynamics and visual scene delays.

a **Objectives**: The main objective is to check the displays utility in allowing the pilot to use the display to during maneuvering at reduced and elevated load factors. A second objective is to check the short term dynamics of the display/aircraft during aggressive aimpoint tracking including ensuring that there are no undesirable cross-coupling between axes.

b **Requirements**: This evaluation requires a cooperative target aircraft.

c **Action**: The maneuver is initiated from level unaccelerated flight with both aircraft at a constant airspeed equal to the V_C of the test aircraft (corner speed). The test aircraft is positioned 2000 ft in trail behind the target aircraft. The target aircraft then decelerates 50 knots to (V_C -50), causing the test aircraft to close on the target. When the range between the two aircraft decreases to approximately 1000 ft, the target aircraft initiates a 5 g banked turn at constant altitude. The test aircraft delays until the line-of-sight reaches 30 deg, at which time the pilot initiates a climbing turn toward the target, with a nose-up pitch attitude of 15 to 30 deg. The resulting deceleration causes a decrease in the rate of closure from above. When the closure rate is no longer apparent, the test aircraft rapidly lowers the nose to achieve a firing solution within weapon system constraints.

d **Data**: Aircraft performance data should be recorded as functions of time:

> Airspeed
> Altitude
> Pitch and bank angles.

Target acquisition and tracking accuracy shall be recorded, preferably using visual image recording. Timing begins as the test aircraft lowers the nose to achieve a firing solution.

Pilot ratings (DRRs and DFRs) should be hand recorded. Voice recording may be used to supplement the hand recorded data.

e **Performance criteria**:

i *Desired performance*:
- Achieve aimpoint within 4 sec within firing constrains of the weapon system to be deployed on the aircraft.
- Acquire the target with no tendency for pitch overshoots.
- No undesirable cross-coupling.

ii *Adequate performance*:
- Achieve aimpoint within 8 sec within firing constrains of the weapon system to be deployed on the aircraft.
- Acquire the target with no tendency for pitch overshoots.
- No objectionable cross-coupling.

f **Safety considerations**: These tests are aggressive maneuvers flown in proximity to other aircraft. Suitable precautions for formation flight should be observed, including pre-test briefings.

Yo-Yo, Low

The low yo-yo is a maneuver used in air-to-air combat tactics. For HMD testing, the aiming reticle on the HMD should be kept on the other aircraft throughout the test. This task is inappropriate for simulator evaluation because of possible interactions between display dynamics and visual scene delays.

a **Objectives**: The main objective is to check the displays utility in allowing the pilot to use the display to during maneuvering at reduced and elevated load factors. A second objective is to check the short term dynamics of the display/aircraft during aggressive aimpoint tracking including ensuring that there are no undesirable cross-coupling between axes.

b **Requirements**: This evaluation requires a cooperative target aircraft.

c **Action**: The maneuver is initiated from level unaccelerated flight with both aircraft at a constant airspeed equal to the V_C of the test aircraft (corner airspeed). The test aircraft is positioned approximately 200 ft in trail behind the target aircraft. The target aircraft then accelerates 50 knots to $(V_C + 50)$, resulting in an increase in range between the two aircraft. When the range between the two aircraft increases to approximately 1000 ft, the target aircraft initiates a 5 g turn at constant altitude. The test aircraft delays until the line-of-sight reaches 30 deg, at which time the pilot initiates a diving turn toward the target, with a nose-down pitch attitude of 15 to 30 deg. The resulting acceleration causes the test aircraft to begin to close on the target from above. When a rate of closure is apparent, the test aircraft rapidly raises the nose and tracks the target to achieve a firing solution within weapon system constraints.

Timing begins as the test aircraft raises the nose to achieve a firing solution.

d **Data**: Aircraft performance data should be recorded as functions of time:

> Airspeed
> Altitude
> Pitch and bank angles.

Target acquisition and tracking accuracy shall be recorded, preferably using visual image recording.

Pilot ratings (DRRs and DFRs) should be hand recorded. Voice recording may be used to supplement the hand recorded data.

e **Performance criteria**:
  i *Desired performance*:
    • Achieve aimpoint within 4 sec within firing constrains of the weapon system to be deployed on the aircraft.
    • Acquire the target with no tendency for pitch overshoots.

- No undesirable cross-coupling.

ii *Adequate performance*:
  - Achieve aimpoint within 8 sec within firing constrains of the weapon system to be deployed on the aircraft.
  - Acquire the target with no tendency for pitch overshoots.
  - No objectionable cross-coupling.

f **Safety considerations**: These tests are aggressive maneuvers flown in proximity to other aircraft. Suitable precautions for formation flight should be observed, including pre-test briefings.

Precise Mission Task Elements

The following are precise mission task elements (MTEs) for fixed-wing aircraft. In most cases, they were adapted from other test and evaluation documents and modified to examine display characteristics and display/aircraft interactions, rather than aircraft characteristics alone. In some cases, the flight in the clear-visual-day environment will seem trivial; however, such evaluations will form the basis for subsequent night or adverse weather evaluations.

Instrument Maneuvers

These maneuvers are standard instrument training maneuvers. These tasks are flown to evaluate the aircraft display symbology and dynamics during tracking. Instrument approaches and terminal area maneuvering will evaluate the cognitive aspects of the display (chapter 13, page 209).

a **Objectives**: This flight test includes a series of basic maneuvers designed to evaluate the display symbology and to demonstrate that maneuvers can be performed within the performance tolerances appropriate to the maneuvers.

b **Requirements**: These test should be flown in appropriate aircraft configurations. These tests should normally be flown in actual or simulated IMC.

c **Action**: The representative instrument tasks from the following list should be flown using normal procedures from instrument training manuals(*139-140*).
  i *vertical S-A*

    ii *vertical S-C*
    iii *vertical S-D*
    iv *oboe patterns*
    v *yankee patterns*
    vi *steep turns* (bank angle depends on aircraft type, through 360 deg, immediate reversal 360 deg back to original heading).

d **Data**: Aircraft performance data should be recorded as functions of time:

> Airspeed
> Barometric altitude
> Radar altitude
> Heading
> Pitch and bank angles.

    Pilot ratings (DRRs and DFRs) should be hand recorded. Voice recording may be used to supplement the hand recorded data.

e **Performance criteria**:
    i *Desired performance*:
- Maintain altitude within ±100 ft during level legs; ±200 ft during level-offs.
- Maintain heading ±3 deg during straight legs; ±5 deg rolling out.
- Maintain airspeed ±5 knots.
- Complete all legs within ±5 sec of standard times.
- Maintain bank (steep turns) within ±3 deg.
- Maintain airspeed (steep turns) ±5 knots.

    ii *Adequate performance*:
- Maintain altitude within ±200 ft during level legs; ±300 ft during level-offs.
- Maintain heading ±5 deg during straight legs; ±10 deg rolling out.
- Maintain airspeed ±5 knots.
- Complete all legs within ±10 sec of standard times.
- Maintain bank (steep turns) within ±5 deg.
- Maintain airspeed (steep turns) ±10 knots.

f **Safety considerations**: These tests involve a hooded evaluation pilot. Adequate scan of the external visual scene by the safety pilot is essential. Care must be taken to ensure that the vision restriction for the evaluation pilot does not impair the vision of the safety pilot.

Instrument Tracking

The evaluations under this MTE are designed solely to evaluate the ability of the pilot to use the display under tracking conditions. Instrument approaches will be evaluated more completely under mission related tasks that will include evaluation of situation awareness, cognitive processing, and other mission tasks.

a **Objectives**: This flight test includes a series of basic maneuvers designed to evaluate the display symbology and to check for pilot-aircraft-display dynamic issues arising during tight compensatory tracking tasks.

b **Requirements**: These test should be flown in appropriate aircraft configurations. These tests should normally be flown in actual or simulated IMC.

c **Action**: An appropriate radio navigation station (VOR, TACAN, or ADF, as appropriate for the installed equipment) should be tuned. From a 45 deg intercept leg, intercept the selected courses. Track both inbound and outbound while maintaining altitude. The tracking should include some station passages.

Tune an ILS facility (if an appropriate receiver is installed). From a 45 deg intercept leg, intercept the selected course. From inbound tracking in the ILS localizer front course, intercept the glideslope and track down toward the runway. Backcourse localizer courses should also be flown.

d **Data**: Aircraft performance data should be recorded as functions of time:

> Airspeed
> Barometric altitude
> Radar altitude
> Heading
> Course deviation
> Glideslope deviation
> Marker beacon signal.

Pilot ratings (DRRs and DFRs) should be hand recorded. Voice recording may be used to supplement the hand recorded data.

e **Performance criteria**:
  i *Desired performance*:
  - Maintain altitude within ±100 ft during level legs.
  - Maintain heading (during intercepts) ±3 deg.

- Maintain airspeed ±5 knots.
- Tracking within ±1/2 dot for fifty percent of tracking task.
- Tracking within ±1 dot for entire of tracking task.
- No overshoots (beyond 1/2 dot) on course intercept.

ii *Adequate performance*:

- Maintain altitude within ±200 ft during level legs.
- Maintain heading (during intercepts) ±5 deg.
- Maintain airspeed ±5 knots.
- Tracking within ±1 dot for fifty percent of tracking task.
- Tracking within ±2 dot for entire of tracking task.
- Single overshoot on course intercept.

f **Safety considerations**: These tests involve a hooded evaluation pilot. Adequate scan of the external visual scene by the safety pilot is essential. Care must be taken to ensure that the vision restriction for the evaluation pilot does not impair the vision of the safety pilot.

13 Mission or Operational Test and Evaluation

The following are mission related task to be flown following MTEs. The MTEs are designed to ensure that the various building blocks of maneuvers used during operations fit with the display and aircraft characteristics. The mission related tasks put these building blocks together to form sequences of tasks that will be flown in an operational setting. This increases the cognitive demands on the pilot. The assessment of these increased cognitive demands is a major objective of these evaluations.

Generally, these tasks will be developed from the intended mission for the aircraft/display being tested. It is not possible to list all possible mission related tasks for all possible aircraft. The following are some representative tasks.

During the course of these evaluations, embedded tasks will be included at unexpected (to the evaluation pilot) times. These embedded tasks will include engine and system failures, unplanned diversions and threats. Sufficient numbers of such embedded tasks should be included to allow for some statistical validity. However, not too many tasks should be added to defeat the concept of unexpected failures and diversions.

Military and Civil Aircraft

Instrument Approaches

These evaluations build on the instrument tracking basic instrument tasks described in the previous chapters. The tracking MTEs concentrated on the gains and control laws required to intercept and track the guidance commands. The present evaluations will add the cognitive and situation awareness (SA) aspects of the display, including timing, orientation, and missed approach decisions.

a **Objectives**: In general the objective is to evaluate the display's ability to provide situation awareness information while performing instrument approaches. Secondarily, the objective is to evaluate the display sym-

209

bology and demonstrate that approaches can be performed within the performance tolerances required.

b **Requirements**: These test should be flown in appropriate aircraft configurations. Aircraft configuration changes should be accomplished according to published or interim aircraft manuals. These tests should normally be flown in simulated IMC.

c **Action**: The representative instrument tasks from the following list shall be flown using published procedures.

- Precision approaches shall be flown using normal procedures.
- Non-precision approaches shall be flown using normal procedures. Both front course and back course localizer approaches should be flown.
- For rotary-wing aircraft, Decelerating approaches, described in paragraph 4.3 of ADS-33 (*137*) should be flown unless they are to be prohibited by operating limitations.
- Other approaches, as appropriate, shall be flown using normal procedures.

If an approach is not to be used operationally, that type of approach need not be flown during these tests.

At least fifty approaches should be flown for the primary approach method (normally ILS approaches) and at least twenty should be flown for secondary approaches. Ninety percent of the approaches should be successful approaches where the aircraft is in a position to land at arrival at minimums.

If factors external to the aircraft or display cause the aircraft to be outside the landing window for a particular approach, such as maneuvering to avoid traffic, that approach will not be considered as an unsuccessful approach.

Half should be flown to touchdown and half to missed approaches. Approaches to at least three different runways should be flown for each type of approach. For multiengine aircraft, at least two approaches of each type should be flown with a simulated engine failure during the approach.

d **Data**: Aircraft performance data should be recorded as functions of time:

> Airspeed
> Barometric altitude
> Radar altitude

>Heading
>Course deviation
>Glideslope deviation
>Marker beacon signal
>Landing gear position
>Flap position.

Pilot ratings (DRRs and DFRs) should be hand recorded. Voice recording may be used to supplement the hand recorded data.

It is important that the evaluation pilot record his assessment of performance on each approach to allow for an assessment of subjective performance assessment.

e **Performance criteria:** Performance criteria shall be obtained from instrument handbooks (*139-140*) or equivalent documents.

  i *Desired performance:*
  - Maintain altitude ±100 ft during level legs.
  - Maintain heading (during intercepts) ±3 deg.
  - Maintain airspeed ±5 knots.
  - Tracking ±1/2 dot at minimums.
  - Tracking ±1/2 dot for fifty percent of final approach.
  - Tracking ±1 dot for entire final approach.
  - No overshoots (beyond 1/2 dot) on course intercept.
  - Missed approach +20/-0 ft of DH for ILS approach.
  - Missed approach ± 5 sec of MAP for non-precision approach.
  - Level-off at +20/-0 ft of MDA for non-precision approach.

  ii *Adequate performance:*
  - Maintain altitude ±200 ft during level legs.
  - Maintain heading (during intercepts) ±5 deg.
  - Maintain airspeed ±5 knots.
  - Tracking ±1 dot at minimums.
  - Tracking ±1 dot for fifty percent of final approach.
  - Tracking ±2 dot for entire final approach.
  - Single overshoot on course intercept.
  - Missed approach +40/-20 ft of DH for ILS approach.
  - Missed approach ± 10 sec of MAP for non-precision approach.
  - Level-off at +40/-20 ft of MDA for non-precision approach.

f **Safety considerations:** These tests involve a hooded evaluation pilot. Adequate scan of the external visual scene by the safety pilot is essential. Care must be taken to ensure that the vision restriction for the evaluation pilot does not impair the vision of the safety pilot.

Terminal Area Maneuvering

These evaluations deal with the integration of the basic instrument tasks described in the previous chapters and in the previous section and the realistic ATC environment. It is emphasized that these are not sterile procedures, but must be conducted in a complete terminal environment with distractions and embedded situation awareness evaluations. To quote from AC-29-2C:

> .... new rotorcraft undergoing IFR certification should be flown in the air traffic control system in actual day and night instrument meteorological conditions.(60)

a **Objectives**: The first objective is to evaluate the display's ability to provide all necessary situation awareness data while performing instrument approaches and terminal area maneuvering in a realistic ATC environment. The second objective is to evaluate the display symbology and to demonstrate that approaches can be performed within the performance tolerances appropriate to the approaches flown.

b **Requirements**: These test should be flown in appropriate aircraft configurations. Aircraft configuration changes will be accomplished according to aircraft manuals or equivalent document. These tests should normally be flown in simulated IMC.

c **Action**: The aircraft will be flown from facility (airport, heliport, or other operating location) to facility. Normal ATC procedures to depart the terminal area and enter the arrival terminal area will be followed. Representative terminal areas will be chosen to encompass the range of operating environments expected:. These should range from non-radar approach/departure control through Class B terminal areas. Facilities should range from non-tower locations through full tower/approach control airports. Shipboard operations should be included if appropriate. At lease three levels of ATC control/traffic density should be flown.

Instrument approaches should be flown:
- Precision approaches.
- Non-precision approaches to both tower-controlled and non-tower facilities.

- Visual approaches to both tower controlled and non-tower facilities.

It is expected that facilities not flown will be reflected in operational limitations. If an approach is not to be used operationally, that type of approach need not be flown during these tests.

d **Data**: Aircraft performance data should be recorded electronically, preferably from the display databus. Pilot ratings (DRRs and DFRs) should be hand-recorded. Voice recording may be used to supplement the hand recorded data.

e **Performance criteria**: Performance criteria shall be obtained from reference (*139*) or equivalent document. General IFR criteria are listed.

   i *Desired performance*:
- Maintain altitude ±100 ft during level legs.
- Maintain heading ±3 deg.
- Maintain airspeed ±5 knots.
- Tracking ±1 dot.
- Other criteria as appropriate.

   ii *Adequate performance*:
- Maintain altitude ±200 ft during level legs.
- Maintain heading ±5 deg.
- Maintain airspeed ±5 knots.
- Tracking ±1.5 dots.
- Other criteria as appropriate.

f **Safety considerations**: These tests involve flight in high density traffic or in areas of none or limited air traffic services. Vigilance for other traffic is imperative. For those tests involving a hooded evaluation pilot, adequate scan of the external visual scene by the safety pilot is essential. Care must be taken to ensure that the vision restriction for the evaluation pilot does not impair the vision of the safety pilot.

Training Assessment

Past display systems have required excessive training requirements. The AH-64, *Apache*, requires approximately twelve hours of checkout. The IHADSS helmet-mounted display requires approximately fifty hours to learn to use.(*162*) For this reason, the draft *HMD Aeronautical Design Standard* (*113*) recommends that a life-cycle training cost estimate be provided as a deliverable report during military procurements. It also recommends an assessment of training be conducted during the HMD test and evaluation.

Training of display evaluators is an important issue (see page 98). Haworth and Newman (*23*) discussed this in some detail.* To some extent, the need do ensure that evaluation pilots are adequately training in a particular display drives the choice of what kind and how many evaluation pilots to use. Since we have argued for the extensive use of operational pilots in conducting mission tasks, we can use this opportunity to conduct an assessment of the training required during the life cycle of the display system.

The development of the training begins, like the design, with the establishment of the objectives and requirements (*vide supra*, page 25). Prior to conducting mission task element (MTE) or mission task testing with operational pilots, a training syllabus is required. If the system is an new aircraft, this training should be typical new-aircraft checkout. If the system is a new display for an existing aircraft, the syllabus may be shortened appropriately. The syllabus should be that on which the life-cycle training cost is based.

The syllabus should include graduation standards in order to determine that the pilots are properly trained.

a   **Objectives**: The objective is to record the training required for representative operational pilots to become qualified on the display or aircraft in questioned.

b   **Requirements**: The training syllabus must be established prior to using non-test pilots in the display or systems evaluation.

c   **Action**: The prospective evaluation pilots should receive training to proficiency. The use of simulators is allowed.

  The pilot population should cover the range of experience expected in operations.

d   **Data**: The training times and final checkride grades should be recorded.

e   **Performance criteria**: There is no safety-related performance criteria. There is a performance-related performance criteria.

---

* We should mention that it is not unusual in civil flight testing for Authority pilots to arrive at the flight test facility and fly evaluations with minimal opportunity for training or familiarization. In our opinion, such practice is inappropriate.

    i *Desired performance*:   Average training is less than predicted times.

    ii *Adequate performance*   Average training matches predicted training times.

f **Safety considerations**: There are no particular safety issues associated with this test.

Other Mission Tasks

The specific tasks and performance criteria for other mission tasks should be obtained from the Aircrew Training Manual for the aircraft or from an equivalent document. These tasks would include such tasks as

- Sling operations (helicopters)
- Platform operations (helicopters)
- Emergency Medical Service (helicopters)
- Low-level patterns (aerial application).

a **Objectives**: These flight tests will demonstrate that mission-specific tasks can be accomplished within the performance tolerances appropriate to the maneuvers.

b **Requirements**: The requirements will be developed for each mission task based on the ATM or equivalent document.

c **Action**: The test procedure will be developed for each mission task based on the ATM or equivalent document.

d **Data**: The data requirements and post-test analyses will be developed for each mission task based on the ATM or equivalent document.

e **Performance criteria**: The requirements will be developed for each mission task based on the ATM or equivalent document. In general, the adequate performance criteria should be identical to the performance criteria stated in the ATM as satisfactory performance. The desired performance criteria should be approximately 1/2 to 2/3 of this value.* The specific values should be agreed to by the customer/user organization.

---

* Such as, maintain airspeed within 10 knots (ATM performance criterion) should translate to adequate performance = ±10 knots and desired performance = ±5 to 7 knots.

Example: proposed sling operations performance criteria, developed from typical hover ATM criteria could be

i   *Desired performance*:   Heading ±5 deg

                                  Altitude ±5 ft

                                  Position ±5 ft.

ii *Adequate performance*   Heading ±10 deg

                                  Altitude ±10 ft

                                  Position ±10 ft.

f **Safety considerations**: Safety considerations must be developed for each aircraft/mission combination.

Military Aircraft

Weapon Delivery Ingress/Egress

These evaluations deal with the integration of the basic flight tasks described in the previous two chapters and a realistic target environment. It is emphasized that these are not sterile procedures, but must be conducted in a complete target environment with distractions and embedded situation awareness evaluations (described in the following Chapter).

a **Objectives**:
- Evaluate the display's ability to provide all necessary situation awareness data while performing target area ingress, weapon delivery, and egress in a realistic threat environment.
- Evaluate the display symbology and to demonstrate that weapon deliveries can be performed within the performance tolerances required.

b **Requirements**: A suitable weapon range and ingress/egress routes are required. Live weapon firing are not required for these evaluations. Meteorological conditions (i. e. night, restricted visibility, etc.) will be flown as specified in the aircraft design mission.

c **Action**: The aircraft will be flown from an enroute point through an ingress route, to an initial point (IP), perform actual or simulated weapon delivery, and via an egress route to an enroute point. All standard operating procedures will be flown according to aircraft manuals.

Allied aircraft and ground forces will be incorporated as required by the aircraft mission.

Suitable threats or other distractions will be embedded in these tasks. Situation awareness (SA) tests will likewise be embedded. It is expected that mission tasks or meteorological conditions not flown should be reflected in operational limitations.

d **Data**: Aircraft performance data should be recorded electronically, preferably from the display databus. Pilot ratings (DRRs and DFRs) should be hand-recorded. Voice recording may be used to supplement the hand recorded data.

e **Performance criteria**: The requirements will be developed for each mission task based on the ATM or equivalent document. In general, the adequate performance criteria should be identical to the performance criteria stated in the ATM as satisfactory performance. The desired performance criteria should be approximately 1/2 to 2/3 of this value. The specific values should be agreed to by the user organizations. (See the footnote on page 215.)

f **Safety considerations**: These tests may involve high workload flight in high density traffic and/or at low levels. Standard range safety precautions must be taken. Vigilance for other traffic is imperative.

Other Mission Tasks

The specific tasks and performance criteria for other mission tasks should be obtained from the Aircrew Training Manual (ATM) for the aircraft or from an equivalent document. These include such tasks as

- Sling operations
- Shipboard operations
- Formation (helicopters and transports)*
- Low-level (ASW) patterns
- High-level (SAR or AWACS) patterns.

a **Objectives**: These flight tests will demonstrate that mission-specific tasks can be accomplished within the performance tolerances appropriate to the maneuvers.

---

* In the context of these types of aircraft, formation can better be expressed as stationkeeping, rather than true formation. These tests are more an evaluation of stationkeeping equipment, such as SKE, rather than formation which is characteristic of fighter and attack aircraft.

b **Requirements**: The requirements will be developed for each mission task based on the ATM or equivalent document.

c **Action**: The test procedure will be developed for each mission task based on the ATM or equivalent document.

d **Data**: The data requirements and post-test analyses will be developed for each mission task based on the ATM or equivalent document.

e **Performance criteria**: The requirements will be developed for each mission task based on the ATM or equivalent document. In general, the acceptable performance criteria should be identical to the performance criteria stated in the ATM as satisfactory performance. The desirable performance criteria should be approximately 1/2 to 2/3 of this value. The specific values should be agreed to by the user organization.

f **Safety considerations**: Safety considerations must be developed for each aircraft/mission combination.

14 Situation Awareness Testing

This chapter describes several flight test techniques for assessing the situation awareness aspects of cockpit flight displays. Background for these discussions is presented in Chapter 7 (page 53).

Direct Evaluations

Attitude Awareness

This flight test complements the standard unusual attitude recovery where the pilot is placed in a UA and tasked to recover. This test is designed to evaluate the ability of the display symbology to convey spatial orientation awareness to the pilot.

a **Objectives**: The objectives are to evaluate the ability of the display and symbology to convey spatial orientation awareness to the pilot and to evaluate the display to provide cues to allow pilots to recognize and recover from inadvertent UAs.

b **Requirements**: These test should be flown in ground based simulators in appropriate aircraft configurations. These tests should be performed in all display modes, including declutter levels.

c **Action**: The evaluation pilot shall be tasked to follow another aircraft in trail formation. The lead aircraft or symbol will fly a series of maneuvers within the capability of the aircraft under evaluation similar to the method of Gallimore *et al. (114)*

The evaluation pilot will be instructed to follow the aircraft, but not to exceed normal limitations in bank and pitch angle, minimum and maximum airspeed, and minimum altitude. These values should be clearly defined. The lead aircraft will, during the course of the test, approach these limits without exceeding them as well as fly beyond them.

When any limit is exceeded, the pilot will be instructed to make a verbal call and recover to straight and level flight using standard procedures from instrument flight manuals (*139-140*) or other published standard. Following recovery, she shall complete displays ratings (DRR and DFR).

If the evaluation pilots fails to note penetration of the envelope limits, the lead aircraft will be removed from the scene after two seconds forcing her to initiate an unusual attitude recovery.

Several replications of these flight envelope limit data points will be required.

d **Data**: Aircraft performance data should be recorded as functions of time:

> Airspeed
> Altitude
> Pitch and bank angles.

Target acquisition and tracking accuracy shall be recorded, preferably using visual image recording.

Pilot ratings (DRRs and DFRs) should be hand recorded. Voice recording may be used to supplement the hand recorded data.

e **Performance criteria**: Performance criteria are the same as unusual attitude recoveries.

i *Desired performance*:

- Abandoned pursuit at aircraft limitations (±5 knot, ±10 deg bank tolerance).
- Abandoned pursuit at altitude floor (±100 ft tolerance).
- Reaction time to initial correct control input < 1.5 sec.
- No initial control reversals.
- Initial control input in accordance with (IAW) published standards. Published standards include instrument flight manuals (*139-140*) aircraft-specific manuals, or equivalent documents.
- No overshoot on recovery.

ii *Adequate performance*:

- Abandoned pursuit at aircraft limitations (±10 knot, ±20 deg bank tolerance).
- Abandoned pursuit at altitude floor (±500 ft tolerance).
- Reaction time to initial correct control input < 2.0 sec.
- Single initial control reversals.

- Initial control input IAW published standards. Published standards include instrument flight manuals (*139-140*) aircraft-specific manuals, or equivalent documents.
- Single overshoot on recovery.

f **Safety considerations**: These tests are flown in the simulator because of the hazards involved with aggressive formation flight in IMC.

Embedded Evaluations

The following are proposed situation awareness (SA) test suitable for display evaluation. In some cases, these have been used in previous programs. There are a number of SA evaluation methods used by psychologists, most of which seek subjective assessment of SA.(*88*) The tests proposed here are based on the presumption that the best measure of SA is the effect on performance. SAP-like assessments are used as well.

Failure Detection:

a **Objectives**: These tests are designed to verify that the pilot can recognize critical failures and take appropriate corrective action while maintaining aircraft control.

b **Requirements**: These test shall be flown in conjunction with other evaluations, particularly the mission task testing.

Navigation, sensor, and display system failures shall be introduced during mission related tasks to determine the reaction time and responses of the pilot. Appropriate procedures for operating the aircraft following these failures shall be established prior to conducting these tests.

It may be required to use a special software version to allow the flight test engineer to introduce some system failures such as simulating attitude gyro failures. If such software versions are used, the response of the system to the simulated failure shall be verified prior to conducting these tests.

Typical failures (and representative simulation means) should include

- Radar altitude failure (pull circuit breaker, C/B)
- Navigation failure (pull C/B)

- Attitude gyro failure (software simulation of both hardover and slow failures)
- Flight director failure (software simulation of both hardover and slow failures)
- Compass failure (pull C/B or software simulation)
- Airspeed or altitude transducer (pull C/B or equivalent)
- Head tracker failure (pull C/B or software simulation).

c **Action**: For tests involving operations in IMC, the evaluation pilot shall be hooded or otherwise prevented from using outside visual cues. System failures shall be introduced during operations at unexpected times by the safety pilot or flight test engineer.

Upon recognition of the failure, the evaluation pilot shall take appropriate action as described in the aircraft/system operating procedures. The evaluation pilot will be briefed to make a verbal call upon recognition of all system failures as a means of determining detection time. Following recovery, the evaluation pilot shall complete displays ratings (DRR and DFR).

d **Data**: Aircraft performance data should be recorded as functions of time:

> Airspeed
> Altitude
> Pitch and bank angles
> Control positions
> Parameters appropriate to failure
> Event (for start of maneuver).

Pilot ratings (DRRs and DFRs) should be hand recorded. Voice recording may be used to indicate the start of the UA and to used to supplement the hand recorded data.

e **Performance criteria**:

i *Desired performance*:

for attitude failures
- Failure detection time within 1 sec.
- Maintain altitude within ±200 ft..
- Maintain airspeed within ±10 knots.
- Maintain heading within ±10 deg.

for heading failures
- Failure detection time within 2 sec.

- Maintain altitude within ±100 ft.
- Maintain airspeed within ±5 knots.
- Maintain heading within ±10 deg.

for other failures
- Failure detection time within TBD.
- Maintain altitude within ±100 ft.
- Maintain airspeed within ±5 knots.
- Maintain heading within ±5 deg.
- Course deviations < 1 dot.

for glideslope failures above localizer MDA
- Minimum altitude = MDA +20/-0 ft.

Note: Slowly developing failures, such as autopilot softovers may require time to detect. In such cases, it is better to describe performance in terms of flight path deviations rather than time.

- Pilot actions in accordance with (IAW) published standards. Published standards include instrument manuals,(*139-140*) air-craft-specific manuals, or equivalent documents.

ii *Adequate performance*:
for attitude failures
- Failure detection time within 2 sec.
- Maintain altitude within ±500 ft.
- Maintain airspeed within ±20 knots.
- Maintain heading within ±20 deg.

for heading failures
- Failure detection time within 5 sec.
- Maintain altitude within ±200 ft.
- Maintain airspeed within ±10 knots.
- Maintain heading within ±20 deg.

for other failures
- Failure detection time within TBD.
- Maintain altitude within ±200 ft.
- Maintain airspeed within ±10 knots.

- Maintain heading within ±10 deg.
- Course deviations < 2 dots.

for glideslope failures above localizer MDA
- Minimum altitude = MDA +40/-20 ft.

Note: Slowly developing failures, such as autopilot softovers may require time to detect. In such cases, it is better to describe performance in terms of flight path deviations rather than time.

- Pilot actions IAW published standards. Published standards include instrument manuals(*139-140*), aircraft-specific manuals, or equivalent documents.

f **Safety considerations**: These tests involve simulated failures. Suitable precautions to ensure the safety pilot retains adequate instrument references and that ATC clearances are not jeopardized.

Geographic Awareness

The recommended scenarios duplicate situations, encountered during normal flight operations where the pilot must maintain geographical orientation.

Typical scenarios should include
- Deviation from a planned flight plan caused by intentional deviations around weather, obstacles, or threats, followed by a return to the original flight plan.
- Deviation from a planned flight plan caused by intentional deviations around weather, obstacles, or threats, followed by a modification to the flight plan (such as bypassing the next waypoint).
- Changing the flight plan route.
- Changing to another, unplanned instrument approach.
- Diverting to an alternate.

These tests are intended to determine if the pilotage display permits the maintenance of geographical awareness. At the same time, other parallel evaluations should be conducted to ensure that the navigation display is suitable and the pilot can enter the necessary data to generate a new flight plan during those tests which involve flight plan modification.

a **Objectives:** These tests are designed to verify that the pilot can maintain geographical situation awareness, determine her orientation relative to desired course, and determine bearing and time to the next waypoint.

b **Requirements:** These test shall be flown in conjunction with other evaluations, particularly the mission simulations conducted under Chapter 13.

c **Action:** The pilot shall be given a typical mission profile to fly. During the course of flying this profile, he will be given diversion instructions by the safety pilot. The diversions should be of the following types:

- A series of heading instructions around a hypothetical weather build-up, an obstacle, or a threat. The evaluation pilot will then be told to return to his original flight plan route.
- A series of heading instructions around a hypothetical weather build-up, an obstacle, or a threat. The evaluation pilot will then be told to fly directly to a waypoint on her original flight plan. This need not be the next sequential waypoint; some evaluations should involve flying to another waypoint..

  These instructions should include terminal area navigation simulating ATC radar vectoring followed by a clearance to join the published instrument approach procedure, either a planned procedure or an alternate procedure.

- An instruction to fly directly to the a subsequent waypoint, bypassing the active waypoint.
- An instruction to divert to an alternate destination.

For tests involving operations in IMC, the evaluation pilot shall be hooded or otherwise prevented from using outside visual cues.

Following recovery, the evaluation pilot shall complete displays ratings (DRRs and DFRs).

d **Data:** Aircraft performance data should be recorded as functions of time:

> Airspeed
> Barometric altitude
> Radar altitude
> Heading
> Course deviation
> Glideslope deviation.

Pilot ratings (DRRs and DFRs) should be hand recorded. It is important to ensure, for the purposes of evaluating the pilotage display, that the evaluators rate the pilotage display not the ease of data entry (although such data entry ratings are important to the overall evaluation of the cockpit). Voice recordings may be used to supplement the hand recorded data.

Pilot estimates of bearing should be obtained from his intitial course decisions. Once established on the new course, the safety pilot/simulator operator should ask for estimates of distances and course line aspect.

e **Performance criteria:**
i *Desired performance*:
- Maintain geographical awareness: enroute
  Bearing to waypoint ± 5 deg
  Distance to waypoint   ± 0.5 nm
  Distance to course line ± 0.2 nm
  Course line aspect   ± 10 deg.
- Maintain geographical awareness: ingress to target
  Bearing to waypoint ± 2 deg
  Distance to waypoint   ± 0.2 nm
  Distance to course line ± 0.1 nm
  Course line aspect   ± 5 deg.
- Maintain geographical awareness: terminal area
  Bearing to waypoint ± 2 deg
  Distance to waypoint   ± 0.2 nm
  Distance to course line ± 0.1 nm
  Course line aspect   ± 5 deg.
- Maintain geographical awareness: instrument approach
  Bearing to waypoint ± 2 deg
  Distance to waypoint   ± 0.1 nm
  Distance to course line ± 0.1 nm
  Course line aspect   ± 2 deg.
- Maintain geographical awareness: during low-level NOE
  Bearing to waypoint ± 5 deg
  Distance to waypoint   ± 600 ft
  Distance to course line ± 300 ft
  Course line aspect   ± 10 deg.
ii *Adequate performance*:
- Maintain geographical awareness: enroute
  Bearing to waypoint ± 10 deg

Distance to waypoint ± 1.0 nm
Distance to course line ± 0.5 nm
Course line aspect ± 30 deg.
- Maintain geographical awareness: ingress to target
Bearing to waypoint ± 5 deg
Distance to waypoint ± 0.5 nm
Distance to course line ± 0.2 nm
Course line aspect ± 10 deg.
- Maintain geographical awareness: terminal area
Bearing to waypoint ± 5 deg
Distance to waypoint ± 0.5 nm
Distance to course line ± 0.2 nm
Course line aspect ± 10 deg.
- Maintain geographical awareness: instrument approach
Bearing to waypoint ± 5 deg
Distance to waypoint ± 0.5 nm
Distance to course line ± 0.2 nm
Course line aspect ± 5 deg.
- Maintain geographical awareness: during low-level NOE
Bearing to waypoint ± 10 deg
Distance to waypoint ± 1200 ft
Distance to course line ± 600 ft
Course line aspect ± 30 deg.

f **Safety considerations**: These tests have no particular hazard other than the high workload for the safety pilot. Standard precautions for instrument training flights should be observed.

Collateral Evaluations

Traffic or Threat Detection

a **Objectives**: These tests are designed to verify that the pilot can maintain adequate awareness of the traffic/threat situation and maintain adequate lookout for threats and other traffic.
b **Requirements**: These tests shall be flown in conjunction with other evaluations.

c **Action**: The evaluation pilot will be briefed to make verbal calls for all observed traffic or threats. Both cooperative aircraft and targets of opportunity may be used.

d **Data**: The safety pilot or observer will make a note of the relative location of the reported traffic/threat at the time the evaluation pilot makes her verbal call. This position should be in relative terms (i. e. one o'clock, two miles). The safety pilot/observer will also note the delay, if any, between his own sighting of the traffic and that call by the evaluation pilot.

e **Performance criteria**: There is no consensus for performance criteria. One civil HUD program compared sighting times between evaluation pilot and safety pilot with acceptable performance being no significant difference in sighting times.

f **Safety considerations**: There are no additional hazards resulting from this evaluation.

Performance Assessment

One issue that was advanced in early HUD certification was the ability of a pilot to assess flight performance.(9) This task has been developed from this issue.

a **Objectives**: Check the ability of the pilot to assess flight performance.

b **Requirements**: These tests shall be flown in conjunction with all evaluations, including MTE testing.

c **Action**: During the course of the evaluations, the pilot will complete display readability and display flyability ratings. These ratings require the pilot to evaluate both his workload and his task performance.

The subjective task performance will be recorded and compared with objective performance data. For example, in a pirouette, the pilot may state his altitude performance as "meeting desired criteria" when, in fact, it fails to meet these criteria.

d **Data**: Aircraft performance and position data should be recorded electronically, preferably from the display databus. Subjective performance estimation shall be obtained from the pilot ratings (DRRs and DFRs).

e **Performance criteria**: Agreement between subjective and objective performance constitutes desirable results. Actual performance better than subjective performance is undesirable. Actual performance worse than subjective performance is unsatisfactory and will be unacceptable if hazardous. The degree of agreement will depend on the aircraft and flight task.

Prior to developing the test plan, agreement must be reached on the degree of agreement required to meet these criteria. In a previous program,(9) an comparison was made of the evaluation pilot's assessment of ILS performance and the actual value. In that program, agreement was deemed satisfactory if the two values were within one dot deviation or five knots in airspeed as that was thought to be the limit of the pilot's recollection of his performance following the approach.

i *Desired performance*:
- Agreement between subjective performance and actual performance within a value dependent on aircraft and flight task.
- Subjective performance assessment conservative.

ii *Adequate performance*:
- Agreement between subjective performance and actual performance within a value dependent on aircraft and flight task.
- Subjective performance assessment non-hazardous.

f **Safety considerations**: There are no additional hazards resulting from this evaluation.

15 Strawman Test Schedules

It is instructive to work through sample flight programs to determine flight test schedule estimates. The following is a hypothetical flight test program. Four examples are presented: Advanced scout helicopter, Fixed-wing fighter, HUD-equipped civil transport, Civil transport (Head-down only).

Advanced Scout Helicopter

Assumptions

The helicopter is assumed to be a twin-engine, fly-by-wire, advanced scout machine. There are two crewmembers: Pilot and copilot/gunner. The minimum crew for non-combat missions is one pilot.

The intended primary flight display is a binocular helmet-mounted display. The pilot station incorporates a standby primary flight display.

The intended mission includes missile weapon delivery and laser designation for cooperating attack aircraft using the HMD. The scout mission is to include night and low-visibility nap-of-the-earth (NOE) operations.

The aircraft is to be fully IFR qualified using ILS, TACAN, and GPS approaches.

The preliminary FMEA shows the following significant failures modes:

i *Failure 1*: Flight control computer failure degrades HQ from Level 1 to Level 2 (MTBF ≈ 10,000 hr).
ii *Failure 2*: Second failure degrades HQ to Level 3.
iii *Failure 3*: AHRS failure degrades HQ from Level 1 to Level 2; simultaneous loss of attitude display on HMD (MTBF ≈ 5,000 hr).
iv *Failure 4*: Second failure degrades HQ to Level 3.
v *Failure 5*: Second type AHRS failure does not degrade HQ; (MTBF ≈ 2,500 hr).
vi *Failure 6*: Second failure degrades HQ to Level 3.

vii *Failure 7*: Head tracker failure freezes world-referenced symbols; (MTBF ≈ 5,000 hr).

Tests to be Conducted

Based on the choices of MTEs shown in Table 8.5, the tests shown in Table 15.1 should be conducted on this helicopter. Table 15.1 shows the test summary and estimated simulator and flight times necessary to conduct these tests.

Table 15.1 Display Test Tasks

Mission Task Element	Grnd	Sim	Flight	Flt Hrs
Ground Tests				
Egress	X			-
Fit and Reach Testing	X			-
Symbol List Testing	X	X		-
Initial Flight Tests				
Mode and Function Evaluation		X	X	6:00
Calibration of Hover Vector		X	X	1:00
Sensor Calibration			X	1:00
TACAN Calibration			X	1:00
Localizer Calibration			X	1:00
Glideslope Calibration			X	1:00
Single-Axis Control Step Inputs		X	X	1:00
Step Inputs to head-tracker (F)		X	X	1:00
Step Inputs to head-tracker (G)	X	X		
				-
Aggressive MTEs				
Accelerations/Deceleration		O	X	1:00
Air-to-Air Visual Tracking		NR	X	1:00
Air-to-Ground Visual Tracking		NR	X	1:00
Bob-Up/Bob-Down		O	X	1:00
Engine Failure		O	X	1:00
Flight Control System Failure		O	X	1:00
Sidestep		O	X	1:00
Slalom		O	X	1:00
Turn-to-Target		O	X	1:00
Turn-to-Target, modified		O	X	1:00
Unusual Attitude Recovery		X	X	2:00
Vertical Remask		O	X	1:00

Note: X = Test mandatory; O = Test optional; E = Embedded in other tests
       R = Recommended for risk reduction; NR = Not recommended

Table 15.1 Display Test Tasks (continued)

Mission Task Element	Grnd	Sim	Flight	Flt Hrs
Precise MTEs				
Hover		O	X	1:00
Hovering Turn		O	X	1:00
Instrument Tasks		O	X	1:00
Instrument Tracking		O	X	2:00
Landing		O	X	1:00
Pirouette		O	X	1:00
Mission Tasks				
Attack Profiles		R	X	2:00
Formation		R	X	1:00
Instrument Approaches		R	X	2:30
Terminal Operations		R	X	2:30
Training Assessment		X		10:00
SA Tasks				
Attitude Awareness		X	NR	2:00
Geographic Awareness		X/E	X/E	2:00
Performance Assessment		X/E	X/E	E
Systems Failure Detection		X/E	X/E	2:00
Traffic or Threat Detection		X/E	X/E	E

Note: X = Test mandatory; O = Test optional; E = Embedded in other tests
R = Recommended for risk reduction; NR = Not recommended

The time estimates in the table are based past experience. One hour was used for each MTE evaluation. This should allow transition to and from the test areas and 4 to 5 replications per flight. An estimate of two hours was made for unusual attitude recoveries and instrument tracking tasks. The figures for mission tasks are based on typical mission durations. The situation awareness values are estimated additions to the tasks in which they are embedded. The estimates for other tasks was based on fixed-wing test experience.

FCS failure tests: Based on the assumptions made above, the flight control systems (FCS) failures tests to be performed during MTE evaluations should include the following tests to determine if the pilot can cope with these failures.

    i   *Failure 1*: FCS computer failure degrades HQ from Level 1 to Level 2.

    ii *Failure 2*: Second FCS failure degrades HQ to Level 3.

iii *Failure 3*: AHRS failure degrades HQ from Level 1 to Level 2; simultaneous loss of attitude display on HMD.

iv *Failure 4*: Second AHRS failure degrades HQ to Level 3.

v *Failures 5 and 6*: Second type AHRS failure (double failure) degrades HQ to Level 3.

Systems failures used in SA tests: Based on the assumptions made above, the failure detection tests to be performed during SA testing should include the following tests to determine if the pilot can detect *and* cope with these failures.

i *Failure 3*: AHRS failure degrades HQ from Level 1 to Level 2 and simultaneous loss of attitude display on HMD.

ii *Failure 5*: Second type AHRS failure does not degrade HQ; (to evaluate ability to detect latent failure)

iii *Failure 7*: Head tracker failure freezes world-referenced symbols.

iv *Radar altitude failure*

v *Navigation failure*

vi *Attitude gyro failure* (both hardover and slow failures)

vii *Flight director failure* (both hardover and slow failures)

viii *Compass failure*

ix *Airspeed or altitude transducer failure.*

Choice of Pilots

Initial tests:	1 Company test pilot
Aggressive MTEs:	1 Company test pilot
	1 Government test pilot
	4 Operational pilots
Precise MTEs:	1 Company test pilot
	1 Government test pilot
	4 Operational pilots
Mission tasks:	20 Operational pilots (simulator)
	4 Operational pilots (flight)

Flight Hour Estimates

Simulator tests: Table 15.2 summarizes the simulator test requirements. As can be seen from the table, 1072:00 hours of simulator testing is recommended. We could reduce the simulator tests somewhat by minimizing the MTE tests (we need to keep the UA recovery tests fixed). This could reduce the simulator time to 796:00 hours. This, however, would increase program risks.

Flight tests: Table 15.3 summarizes the flight test requirements. As can be seen from the table, 447:00 hours of flight testing is expected during development test and evaluation (DT&E).

In addition to the DT&E flight tests, it is important to remember that approximately 400 hours of operational test and evaluation (OT&E) must be conducted to draw out further problems. In particular, the issue of long duration (over 1:00) using an HMD as a primary flight reference can only be accomplished by flying such missions with operational pilots.

Single-Engine Fighter

Assumptions

The aircraft is assumed to be a single-engine, fly-by-wire, fighter/attack airplane, flown by one pilot.

The intended primary flight display is head-down glass instrument suite. A helmet-mounted display is used for weapon aiming purposes and performs the function of the primary flight display during these portions of flight. The instrument suite incorporates a standby primary flight display.

The intended mission includes missile weapon delivery and laser designation for cooperating attack aircraft using the HMD. The air-to-air mission includes both guns and missile weapons.

The aircraft is to be fully IFR qualified using ILS, TACAN, and GPS approaches.

Table 15.2 Scout Helicopter Simulator Test Requirements

Task	Day	Night	DVE	IMC	Total
Initial Flight Tests					
Hover Vector Calibration	1:00				1:00
Mode and Function	4:00				4:00
Single-Axis Step Inputs					
Head-Tracker Step Inputs	1:00				1:00
Aggressive MTEs					
Accel/Decel	6:00	6:00	6:00	6:00	24:00
Air-to-Air Tracking					
Air-to-Ground Tracking					
Bob-up/Bob-down	6:00	6:00	6:00		18:00
Engine Failures	6:00	6:00	6:00	6:00	24:00
FCS Failures	6:00	6:00	6:00	6:00	24:00
Sidestep	6:00	6:00	6:00	6:00	24:00
Slalom	6:00	6:00	6:00		18:00
Turn-to-Target	6:00	6:00	6:00		18:00
Turn-to-Target, Modified	6:00	6:00	6:00		18:00
UA Recovery				12:00	12:00
Vertical Remask	6:00	6:00	6:00		18:00
Precise MTEs					
Hover	6:00	6:00	6:00		18:00
Hovering Turn	6:00	6:00	6:00		18:00
Instrument Tasks				6:00	6:00
Instrument Tracking				12:00	12:00
Landing	6:00	6:00	6:00		18:00
Pirouette	6:00	6:00	6:00		18:00
Mission Tasks					
Attack Profiles	40:00	40:00	40:00		120:00
Formation	6:00	6:00	6:00		18:00
Instrument Approaches				40:00	40:00
Terminal Operations	20:00	20:00		20:00	40:00
Training Assessment	100:00	50:00		50:00	200:00
SA Tasks					
Attitude Awareness				40:00	40:00
Geographic Awareness	40:00	40:00	40:00	40:00	160:00
Performance Assessment					
Failure Detection	40:00	40:00	40:00	40:00	160:00
Threat Detection					
Total	330:00	274:00	204:00	284:00	1072:00

Table 15.3 Scout Helicopter Flight Test Requirements

Task	Day	Night	DVE	IMC	Total
Initial Flight Tests					
Calibrations	5:00				5:00
Mode & Function	4:00				4:00
Single-Axis Step Inputs	1:00				1:00
Head-Tracker Step Inputs	1:00				1:00
Aggressive MTEs					
Accel/Decel	6:00	6:00	6:00	6:00	24:00
Air-to-Air Tracking	6:00				6:00
Air-to-Ground Tracking	6:00				6:00
Bob-up/Bob-down	6:00	6:00	6:00		18:00
Engine Failure	6:00	6:00	6:00	6:00	24:00
FCS Failure	6:00	6:00	6:00	6:00	24:00
Sidestep	6:00	6:00	6:00	6:00	24:00
Slalom	6:00	6:00	6:00		18:00
Turn-to-Target	6:00	6:00	6:00		18:00
Turn-to-Target, Modified	6:00	6:00	6:00		18:00
UA Recovery (HMD)				12:00	12:00
UA Recovery (S-by)				12:00	12:00
Vertical Remask	6:00	6:00	6:00		18:00
Precise MTEs					
Hover	6:00	6:00	6:00		18:00
Hovering Turn	6:00	6:00	6:00		18:00
Instrument Tasks				6:00	6:00
Instrument Tracking				12:00	12:00
Landing	6:00	6:00	6:00		18:00
Pirouette	6:00	6:00	6:00		18:00
Mission Tasks					
Attack Profiles	8:00	8:00	8:00		24:00
Formation	4:00	4:00	4:00		12:00
Instrument Approaches				8:00	8:00
Terminal Operations	4:00	2:00		2:00	8:00
Training Assessment					
SA Tasks					
Attitude Awareness				8:00	8:00
Geographic Awareness	8:00	8:00	8:00	8:00	32:00
Performance Assessment					
Failure Detection	8:00	8:00	8:00	8:00	32:00
Threat Detection					
Total	133:00	108:00	106:00	100:00	447:00

The preliminary FMEA shows the following significant failures modes:

i _Failure 1_: Flight control computer failure degrades HQ from Level 1 to Level 2 (MTBF ≈ 10,000 hr).

  ii *Failure 2*: Second failure degrades HQ to Level 3.

  iii *Failure 3*: AHRS failure degrades HQ from Level 1 to Level 2; simultaneous loss of attitude display on HMD (MTBF $\approx$ 5,000 hr).

  iv *Failure 4*: Second failure degrades HQ to Level 3.

  v *Failure 5*:   Second type AHRS failure does not degrade HQ (MTBF $\approx$ 2,500 hr).

  vi *Failure 6*: Second failure degrades HQ to Level 3.

  vii *Failure 7*: Head tracker failure freezes world-referenced symbols (MTBF $\approx$ 5,000 hr).

Tests to be Conducted

Based on the choices of MTEs shown in Table 8.6, the following tests should be conducted on this airplane. Because it is a single-engine airplane, no engine failure tests need be flown to evaluate the display. Table 15.4 shows the test summary.

     The time estimates in the table are based past experience. One hour was used for each MTE evaluation. This should allow transition to and from the test areas and 4 to 5 replications per flight. An estimate of two hours was made for unusual attitude recoveries and instrument tracking tasks. The figures for mission tasks are based on typical mission durations. The situation awareness values are estimated additions to the tasks in which they are embedded. The estimates for other tasks was based on fixed-wing test experience.

FCS failure tests: Based on the assumptions made above, the flight control systems (FCS) failures tests to be performed should include the following tests to determine if the pilot can cope with these failures.

  i *Failure 1*: FCS computer failure degrades HQ from Level 1 to Level 2.

  ii *Failure 2*: Second FCS failure degrades HQ to Level 3.

  iii *Failure 3*: AHRS failure degrades HQ from Level 1 to Level 2; simultaneous loss of attitude display on HMD.

  iv *Failure 4*: Second AHRS failure degrades HQ to Level 3.

  v *Failures 5 and 6*: Second type AHRS failure (double failure) degrades HQ to Level 3.

Table 15.4 Fixed-Wing Display Test Tasks

Mission Task Element	Grnd	Sim	Flight	Flt Hrs
Ground Tests				
Egress	X			
Fit and Reach Testing	X			
Symbol List Testing	X	X		
Initial Flight Tests				
Mode and Function		X	X	6:00
Calibrations			X	4:00
Single-Axis Step Inputs		X	X	1:00
Head Tracker Step Inputs	X	X	X	1:00
Aggressive MTEs				
Acrobatics		O	X	1:00
Air-to-Air Tracking		NR	X	1:00
Air-to-Ground Tracking		NR	X	1:00
Engine Failure		O	X	1:00
FCS Failure		O	X	1:00
Formation		O	X	1:00
Refueling		O	X	1:00
Steep Turn and Reversal		O	X	1:00
Turn-to-Target		O	X	1:00
Turn-to-Target, modified		O	X	1:00
UA Recovery		X	X	2:00
Yo-yo, high		O	X	1:00
Yo-yo, low		O	X	1:00
Precise MTEs				
Instrument Tasks		O	X	1:00
Instrument Tracking		O	X	2:00
Mission Tasks				
Attack Profiles		R	X	2:00
Formation		R	X	1:00
Instrument Approaches		R	X	2:30
Terminal Operations		R	X	2:30
Training Assessment		X		10:00
SA Tasks				
Attitude Awareness		X	NR	2:00
Geographic Awareness		X/E	X/E	2:00
Performance Assessment		X/E	X/E	E
Failure Detection		X/E	X/E	2:00
Threat Detection		X/E	X/E	E

Note: X = Test mandatory; O = Test optional; E = Embedded in other tests

R = Recommended for risk reduction; NR = Not recommended

Systems failures used in SA tests: Based on the assumptions made above, the failure detection tests to be performed during SA testing should include the following tests to determine if the pilot can detect *and* cope with these failures.

  i *Failure 3*: AHRS failure degrades HQ from Level 1 to Level 2 and simultaneous loss of attitude display on PFD and HMD.

  ii *Failure 5*: Second type AHRS failure does not degrade HQ; (to evaluate ability to detect latent failure)

  iii *Failure 7*: Head tracker failure freezes world-referenced symbols

  iv *Radar altitude failure*

  v *Navigation failure*

  vi *Attitude gyro failure* (both hardover and slow failures)

  vii *Flight director failure* (both hardover and slow failures)

  viii *Compass failure*

  ix *Airspeed or altitude transducer failure*

Choice of Pilots:

Initial tests:	1 Company test pilot
Aggressive MTEs:	1 Company test pilot 1 Government test pilot 4 Operational pilots
Precise MTEs:	1 Company test pilot 1 Government test pilot 4 Operational pilots
Mission tasks:	20 Operational pilots (simulator) 4 Operational pilots (flight)

Flight Hour Estimates

Simulator tests: Table 15.5 summarizes the simulator test requirements. As can be seen from the table, 967:00 hours of simulator testing is recommended. Some tasks are flown with both head-down displays (HDD) and HMD. This was not a factor in the scout helicopter since the HMD in that aircraft was the only primary flight display. In the fighter, the HMD is a PFD for weapons delivery and the HDD is the PFD for the remainder of flight.

Table 15.5 Single-Engine Fighter Simulator Test Requirements

Task	Day	Night	DVE	IMC	Total
Initial Flight Tests					
Calibrations					
Mode & Function (HDD)	4:00				4:00
Mode & Function (HMD)	4:00				4:00
Single-Axis Step Inputs					
Head Tracker Step Inputs	1:00				1:00
Aggressive MTEs					
Acrobatics (HDD)	6:00	6:00	6:00	6:00	24:00
Acrobatics (HMD)	6:00	6:00	6:00	6:00	24:00
Air-to-Air Tracking					
Air-to-Ground Tracking					
FCS Failures(HDD)	6:00	6:00	6:00	6:00	24:00
FCS Failures(HMD)	6:00	6:00	6:00	6:00	24:00
Formation (HMD)	6:00	6:00	6:00	6:00	24:00
Turn-to-Target	6:00	6:00	6:00		18:00
Turn-to-Target, Modified	6:00	6:00	6:00		18:00
UA Recovery (HDD)				12:00	12:00
UA Recovery (HMD)				12:00	12:00
Yo-Yo, High					
Yo-yo, low					
Precise MTEs					
Instrument Tasks				6:00	6:00
Instrument Tracking				12:00	12:00
Mission Tasks					
Attack Profiles	40:00	40:00	40:00		120:00
Instrument Approaches				40:00	40:00
Terminal Operations	20:00	20:00		20:00	40:00
Training Assessment	100:00	50:00		50:00	200:00
SA Tasks					
Attitude Awareness				40:00	40:00
Geographic Awareness	40:00	40:00	40:00	40:00	160:00
Performance Assessment					
Failure Detection	40:00	40:00	40:00	40:00	160:00
Threat Detection					
Total	291:00	232:00	162:00	302:00	967:00

Flight tests: Table 15.6 summarizes the flight test requirements. As can be seen from the table, 324:00 hours of flight testing is expected during development test and evaluation (DT&E). The unusual attitude recovery tasks can not be flown in the airplane, since only a single-seat airplane is assumed to be available. The UA tasks (amounting to 24:00 hours) should be flown in a variable stability airplane, such as VISTA.(*163*)

HUD-Equipped Transport

Assumptions

The aircraft is assumed to be a HUD-equipped transport airplane with wing-mounted engines. A fly-by-wire control system is assumed.

The intended primary flight display is head-down glass instrument suite. A HUD is incorporated for Category III approaches and will be qualified as an alternate primary flight display for use during other flight phases as well. The instrument suite incorporates a standby primary flight display.

The preliminary FMEA shows the same significant failures modes as in the previous section (again, with the exception of the head-tracker).

Tests to be Conducted

Table 15.4 shows the test summary. The time requirements for category III approaches have been included. Otherwise, the time estimates are the same as in the previous section.

FCS failure tests: Based on the assumptions made above, the flight control systems (FCS) failure tests to be performed should include the following tests to determine if the pilot can cope with these failures.

    i *Failure 1*: FCS computer failure degrades HQ from Level 1 to Level 2.

    ii *Failure 2*: Second FCS failure degrades HQ to Level 3.

    iii *Failure 3*: AHRS failure degrades HQ from Level 1 to Level 2; simultaneous loss of attitude display on the primary flight display (PFD) and the HUD.

    iv *Failure 4*: Second AHRS failure degrades HQ to Level 3.

    v *Failures 5 and 6*: Second type AHRS failure (double failure) degrades HQ to Level 3.

Table 15.6 Single-Engine Fighter Flight Test Requirements

Task	Day	Night	DVE	IMC	Total
Initial Flight Tests					
Mode & Function (HDD)	4:00				4:00
Mode & Function (HMD)	4:00				4:00
Calibrations	4:00				4:00
Single-Axis Step Inputs	1:00				1:00
Head Tracker Step Inputs	1:00				1:00
Aggressive MTEs					
Acrobatics (HDD)	6:00			6:00	12:00
Acrobatics (HMD)	6:00			6:00	12:00
Air-to-Air Tracking	6:00				6:00
Air-to-Ground Tracking	6:00				6:00
FCS Failure (HDD)	6:00	6:00	6:00	6:00	24:00
FCS Failure (HMD)	6:00	6:00	6:00	6:00	24:00
Formation	6:00	6:00	6:00		18:00
Refueling	6:00	6:00	6:00		18:00
Turn-to-Target	6:00	6:00	6:00		18:00
Turn-to-Target, Modified	6:00	6:00	6:00		18:00
MTEs Flown in VISTA					
UA Recovery (HDD)				12:00	12:00
UA Recovery (HMD)				12:00	12:00
Precise MTEs					
Instrument Tasks				6:00	6:00
Instrument Tracking				12:00	12:00
Mission Tasks					
Attack Profiles	8:00	8:00	8:00		24:00
Instrument Approaches				8:00	8:00
Terminal Operations	4:00	2:00		2:00	8:00
Training Assessment					
SA Tasks					
Attitude Awareness				8:00	8:00
Geographic Awareness	8:00	8:00	8:00	8:00	32:00
Performance Assessment					
Failure Detection	8:00	8:00	8:00	8:00	32:00
Threat Detection					
Total	102:00	62:00	60:00	100:00	324:00

Systems failures used in SA tests: Based on the assumptions made above, the Systems Failure Detection tests to be performed should include the following tests to determine if the pilot can detect *and* cope with these failures.

i *Failure 3*: AHRS failure degrades HQ from Level 1 to Level 2 and simultaneous loss of attitude display on both the PFD and the HUD.

ii *Failure 5*: Second type AHRS failure does not degrade HQ (to evaluate ability to detect latent failure).

iii *Radar altitude failure*

iv *Navigation failure*

v *Attitude gyro failure* (both hardover and slow failures)

vi *Flight director failure* (both hardover and slow failures)

vii *Compass failure*

viii *Airspeed or altitude transducer failure.*

Choice of Pilots

Initial tests:	1 Company test pilot
Aggressive MTEs:	1 Company test pilot 1 Authority (FAA, JAA, or CAA) test pilot 4 Operational pilots
Precise MTEs:	1 Company test pilot 1 Authority test pilot 4 Operational pilots
Mission tasks:	20 Operational pilots (simulator) 4 Operational pilots (flight) 6 Operational pilots (ILS approaches)

Flight Hour Estimates

Simulator tests: Table 15.7 summarizes the simulator test requirements. As can be seen from the table, 1060:00 hours of simulator testing is recommended. Because the HUD is an alternate PFD, many of the tests are conducted twice, once head-down and once head-up.

Table 15.7 HUD-Equipped Transport Simulator Test Requirements

Task	Day	Night	DVE	IMC	Total
Initial Flight Tests					
Calibrations					
Mode & Function (HDD)	4:00				4:00
Mode & Function (HUD)	4:00				4:00
Single-Axis Step Inputs					
Aggressive MTEs					
Air-to-Ground Tracking					
Engine Failures (HDD)	6:00	6:00	6:00	6:00	24:00
Engine Failures (HUD)	6:00	6:00	6:00	6:00	24:00
FCS Failures (HDD)	6:00	6:00	6:00	6:00	24:00
FCS Failures (HUD)	6:00	6:00	6:00	6:00	24:00
Steep Turns (HDD)	6:00	6:00	6:00		18:00
Steep Turns (HUD)	6:00	6:00	6:00		18:00
UA Recovery (HDD)				12:00	12:00
UA Recovery (HUD)				12:00	12:00
Precise MTEs					
Instrument Tasks (HDD)				6:00	6:00
Instrument Tasks (HUD)				6:00	6:00
Instrument Track'g (HDD)				12:00	12:00
Instrument Track'g (HUD)				12:00	12:00
Mission Tasks					
Cat III Approaches				200:00	200:00
Instrument Approaches				20:00	20:00
Terminal Ops (HDD)	20:00	20:00		20:00	40:00
Terminal Ops (HUD)	20:00	20:00		20:00	40:00
Training Assessment	100:00	50:00		50:00	200:00
SA Tasks					
Attitude Awareness				40:00	40:00
Geographic Awareness	40:00	40:00	40:00	40:00	160:00
Performance Assessment					
Failure Detection	40:00	40:00	40:00	40:00	160:00
Traffic Detection					
Total	264:00	206:00	116:00	514:00	1060:00

Flight tests: Table 15.8 summarizes the flight test requirements. As can be seen from the table, 419:00 hours of flight testing is expected.

The figures indicated in the previous two paragraphs include 200:00 hours of simulator and 60:00+ hours of flight test devoted to Category III certification. These figures could be subtracted from the totals above if Category III approval were not required. In all likelihood, initial certification would not include Category III; such certification normally follows after the airplane enters service.

If the HUD is an add-on unit for an aircraft already in service, it will only be necessary to complete the portions of the flight test utilizing the HUD. If we subtract the HDD tests from Tables 15.7 and 15.8, only 900 hours of simulator tests and 311:00 hours of flight test would be required.

Glass Cockpit Transport

Assumptions

The aircraft is assumed to be a fly-by-wire transport airplane with wing-mounted engines. The intended primary flight display is head-down glass instrument suite. No HUD is installed. The instrument suite incorporates a standby primary flight display.

The preliminary FMEA shows the same failures modes as the previous case.

Tests to be Conducted

Based on the choices of MTEs shown in Table 8.6, the same tests should be conducted on this aircraft as for the HUD-equipped transport (except that HUD-specific tests, such as aggressive visual tracking, are omitted. Table 15.7 shows the test summary and appropriate visual conditions.

Choice of Pilots

The choices for evaluation pilots should be the same as the previous case.

Table 15.8 HUD-Equipped Transport Flight Test Requirements

Task	Day	Night	DVE	IMC	Total
Initial Flight Tests					
Mode & Function (HDD)	4:00				4:00
Mode & Function (HUD)	4:00				4:00
Calibrations	4:00				4:00
Single-Axis Step Inputs	1:00				1:00
Aggressive MTEs					
Air-to-Ground Tracking	6:00				6:00
Engine Failure (HDD)	6:00	6:00	6:00	6:00	24:00
Engine Failure (HUD)	6:00	6:00	6:00	6:00	24:00
FCS Failure (HDD)	6:00	6:00	6:00	6:00	24:00
FCS Failure (HUD)	6:00	6:00	6:00	6:00	24:00
Steep Turns (HDD)	6:00	6:00		6:00	18:00
Steep Turns (HUD)	6:00	6:00		6:00	18:00
UA Recovery (HDD)				12:00	12:00
UA Recovery (HUD)				12:00	12:00
Precise MTEs					
Instrument Tasks (HDD)				6:00	6:00
Instrument Tasks (HUD)				6:00	6:00
Instrument Track'g (HDD)				12:00	12:00
Instrument Track'g (HUD)				12:00	12:00
Mission Tasks					
Cat III Approaches				60:00	60:00
Instrument Approaches				60:00	60:00
Terminal Ops (HDD)	4:00	2:00		2:00	8:00
Terminal Ops (HUD)	4:00	2:00		2:00	8:00
Training Assessment					
SA Tasks					
Attitude Awareness				8:00	8:00
Geographic Awareness	8:00	8:00	8:00	8:00	32:00
Performance Assessment					
Failure Detection	8:00	8:00	8:00	8:00	32:00
Traffic Detection					
Total	79:00	56:00	40:00	244:00	419:00

Flight Hour Estimates

Simulator tests: Table 15.9 summarizes the simulator test requirements. Nine hundred hours of simulator testing is recommended.

Flight tests: Table 15.10 summarizes the flight test requirements amounting to 245:00 hours.

Category III: The figures indicated in the previous two paragraphs include 200:00 hours of simulator and 60:00 hours of flight test devoted to Category III certification. These figures could be subtracted from the totals above if Category III approval were not required. In all likelihood, initial certification would not include Category III; such certification would follow after the airplane were in service.

Table 15.9 Transport Simulator Test Requirements

Task	Day	Night	DVE	IMC	Total
Initial Flight Tests					
Calibrations					
Mode & Function	4:00				4:00
Single-Axis Step Inputs					
Aggressive MTEs					
Engine Failures	6:00	6:00	6:00	6:00	24:00
FCS Failures	6:00	6:00	6:00	6:00	24:00
Steep Turns	6:00	6:00	6:00		18:00
UA Recovery				12:00	12:00
Precise MTEs					
Instrument Tasks				6:00	6:00
Instrument Tracking				12:00	12:00
Mission Tasks					
Instrument Approaches				200:00	200:00
Terminal Operations	20:00	20:00		20:00	40:00
Training Assessment	100:00	50:00		50:00	200:00
SA Tasks					
Attitude Awareness				40:00	40:00
Geographic Awareness	40:00	40:00	40:00	40:00	160:00
Performance Assessment					
Failure Detection	40:00	40:00	40:00	40:00	160:00
Traffic Detection					
Total	222:00	168:00	98:00	432:00	900:00

Table 15.10 Transport Flight Test Requirements

Task	Day	Night	DVE	IMC	Total
Initial Flight Tests					
Mode & Function	4:00				4:00
Calibrations	4:00				4:00
Single-Axis Step Inputs	1:00				1:00
Aggressive MTEs					
Steep Turns	6:00	6:00		6:00	18:00
Engine Failure	6:00	6:00	6:00	6:00	24:00
FCS Failure	6:00	6:00	6:00	6:00	24:00
UA Recovery				12:00	12:00
Precise MTEs					
Instrument Tasks				6:00	6:00
Instrument Tracking				12:00	12:00
Mission Tasks					
Instrument Approaches				60:00	60:00
Terminal Operations	4:00	2:00		2:00	8:00
Training Assessment					
SA Tasks					
Attitude Awareness				8:00	8:00
Geographic Awareness	8:00	8:00	8:00	8:00	32:00
Performance Assessment					
Failure Detection	8:00	8:00	8:00	8:00	32:00
Traffic Detection					
Total	47:00	36:00	28:00	134:00	245:00

16 Summary

Standardization

There is too much emphasis on symbols and their shapes during HUD evaluations. There has been a marked tendency for evaluators to concentrate on changes to the icons* and not on changes to the content.

Historical Description

Most of the effort on the civil side have concentrated on matching the basic-T cluster of flight instruments (airspeed, attitude, altitude, and heading) plus the ancillary instruments required by FAR 25(*11*). Scant attention has been paid to determining exactly what the pilot's needs are using modern instruments.

This has resulted in requirements to display turn-and-bank indicators in some transports or to show a pitch symbol and a vertical speed readout in a flight path based HUD. We are not stating that these are not required; that is the purpose of the information analysis. We are objecting to requiring instruments simply because they were needed in 1943.

The requirements for a primary flight display implicitly assume that the displays are fixed throughout the flight. This was true in the era of steam gauges, but is no longer true today. There is no longer a reason to insist that all data be displayed all the time. Nor is there any reason to insist on fixed formats.

Need for a New Paradigm

The information displayed in a primary flight display should be based on an information requirements study and may vary from flight task to flight

---

* This has been described as bordering on a religious experience. (Bill Best, quoted in reference (*20*)).

task. If information is not needed during a particular flight task, *it need not be displayed.**

Furthermore, there is no longer a strict need for complete standardization. We propose "standardization by exception." This means that only symbology creating a conflict with population stereotypes or having serious internal conflicts would be disapproved.

This would allow, for example, deviations in the basic T where the heading is above the aircraft reference§, but not deviations where the airspeed is on the right and altitude on the left.

Symbology Non-Standardization

We are spending considerable resources on symbology standardization and certification. These resources would be better spent elsewhere. As a result, we have come to the conclusion that symbology need *not* be standardized.

The present rules practically guarantee a lack of progress as designers seek to duplicate existing displays. We struggle to convince the certification authorities that a change is acceptable. Personal preference today carries more weight than human factors studies. The interested reader should review several recent certification evaluations of vertical tapes versus round dials. We must have a performance basis for display decisions.

The specific data requirements of the airworthiness criteria should be eliminated in favor of the information requirements as determined during the initial portion of the display design process. The authorities should no longer certify symbology.

Market-Driven Symbology

Symbology choices should be made by the manufacturer and customer, not the certification authority. The manufacturer/customer should provide the evaluation pilots who will decide the acceptability of the display. The certification pilots should verify function only and oversee the test and evaluation.

---

* We are not quite ready to state "... it *must* not be displayed."

§ either flight path reference or pitch reference.

The novelty of the display will determine the training requirements for the line pilots. This will minimize the likelihood of extreme changes — no customer (i. e. no operator) will want a display with excessive training requirements.

Standardization

The standardization requirements should be limited. The display should not be incompatible with other standards.

The preparation of an analysis justifying the symbology should be a deliverable document for certification. If, after review of this report, the certification authority pilots feel that the symbology has flaws, a specific test, *with objective data criteria*, should be agreed upon for operational pilots to fly. This is the key role for the certification pilots.

Does this mean that there will be no evaluation of new displays? Of course not. Test and evaluation will always be required.

Recommendations

Need for Adequate Requirements

Any display development program must have the requirements stated explicitly. The requirements, stated in an Operational Requirements Document (ORD) or similar document are required for the designer to address the customer's needs.

In addition, the requirements are necessary to develop the test procedures and success criteria for passing the test program.

Need for Mission Analysis and Information Requirements Study

The first step of any cockpit display design program must be to determine what the tasks required are. This results in the Mission Analysis and Information Requirements (MAIR) document, required by the designer to develop the cockpit systems. Without a MAIR, the designer is groping in the dark.

Need for Objective Test Criteria

The overriding goal of any test plan must be to state all pass/fail criteria in terms of objective performance-based criteria. If one can't state objective performance criteria, then one doesn't fully understand the issues. These criteria should be developed from the requirements document to address the customer's needs.

The Final Authority is the Mission-Based Test

Most testing leading up to mission testing is conducted to reduce program risk. A system can pass all tests up to the mission testing and be unacceptable if it does not pass mission and situation awareness testing.

Related to this is the need to conduct mission and situation awareness testing using operational pilots of varying experience. Test pilots should direct the test, not perform the evaluation of cockpit systems in mission testing. If they have concerns, it is their job to develop scenarios to test their concerns, to be flown by operational pilots.

Need for Documentation

It is important that all design and evaluation decisions and test results be documented. Unfortunately, recent trends in the industry are leading away from good documentation practices as a cost savings practice. This is short-sighted, in our opinion. Table 4.2 lists the documents developed during a typical programme.

Definitions

Airworthiness Release: The approval allowing the aircraft to fly operationally. In civil programs this is normally the issuance of the Type Certificate, although approval for airline operational use will also require an amendment to the particular airlines Operations Specifications.

Authority: The government agency responsible for certifying a civil aircraft design as meeting the airworthiness requirements. In the USA, the *Authority* is the FAA.

Category I: A precision instrument approach and landing with a decision height not lower than 60 m (200 ft) and with either a visibility not less than 800 m (2400 ft), or a runway visual range not less than 550 m (1800 ft).(*69*)

Category II: A precision instrument approach and landing with a decision height lower than 60 m (200 ft) but not lower than 30 m (100 ft) and a runway visual range not less than 350 m (1200 ft).(*69*)

Category III: A precision instrument approach and landing with a decision height lower than 30 m (100 ft), or no decision height.(*69*) It is divided into three sub-categories. *Category IIIa* has a decision height lower than 30 m (100 ft), or no decision height and a runway visual range not less than 200 m (700 ft). *Category IIIb* has a decision height lower than 15 m (50 ft), or no decision height and a runway visual range less than 200 m (700 ft) but not less than 50 m (150 ft). *Category IIIa* has no decision height and no runway visual range limitations.

Certification is the process of documenting that the system meets the external formal requirements, whether contractual requirements or the minimum standards imposed by government organizations. Certification determines that the system, including its documentation, analysis, and test results meet these externally imposed requirements.

Configuration, Equipment: The description of the aircraft and installed systems, including hardware, instrumentation, and software.

Configuration, Operational: The operational state of the aircraft: A configuration is defined by the positions and adjustments of the various selectors and controls available to the pilot, except for longitudinal, lateral, directional, vertical, power, and trim controls. The selected configurations to be examined must consist of those required for performance of the operational missions.

Note: The definition of operational configuration includes display modes.

Crew Station Design Document (CSDD):   The document describing the pilot-vehicle interface in terms of the physical cockpit arrangement, the primary and secondary controls, the installed displays including a description of the various systems modes and, for displays, declutter options.

Design Eye Position (DEP): The point, specified by the airframe manufacturer, from which the pilot can view cockpit instrumentation, have adequate external view, and can reach cockpit controls.(*65*)

Design Eye Reference Point (DERP): The spatial position of the observer's eye relative to the optical axis designated by the HUD manufacturer.(*75*)

Extremely Improbable: For civil aircraft, conditions so unlikely that they are not anticipated to occur during the entire operational life of an aircraft type (fleet).(*43*) For military aircraft, extremely improbable means that the probability of occurrence cannot be distinguished from zero and that it is so unlikely that it can be assumed that this hazard will not be experienced in the entire fleet.(*42*)

For civil aircraft, extremely improbable is generally taken to mean less than once per billion hours.(*43*)*   See *Improbable* and *Extremely Remote*.

Extremely Remote: Conditions so unlikely that they are not anticipated to occur during the entire operational life of the fleet, but cannot be disregarded.(*43*)

For civil aircraft, extremely remote is generally taken to mean less than once per ten million hours.(*164*) See *Extremely Improbable* and *Improbable*.

Flight Release: The formal approval allowing the aircraft to fly in a particular configuration. The flight release may include appropriate limitations on weather, maneuvers, procedures, etc.

Flight Test Release: The formal approval allowing the aircraft to fly in a test configuration.

---

* A billion hours is 114077 years.

Note: Flight test releases are usually issued incrementally. For example, a initial release might restrict the flights to day-VMC flight. Following successful initial testing, the release could be extended to night or instrument flight.

Flight Task: A task, accomplished during flight, that can be treated as an isolated event for the purpose of test, evaluation, and analysis.

Flying Qualities (FQ): The response characteristics of the aircraft to the pilot's controls and to external disturbances. Thus FQ consist of the aircraft stability and control characteristics including the characteristics of the flight controls.

Flying qualities will be different with different parts of the augmentation systems operating. Unlike the handling qualities, FQ do not include the displays, and are not affected by flying at night or in poor visibility.

General Aviation: Civil aviation other than airline operations.

Handling Qualities (HQ): Those characteristics of an aircraft which govern the ease and precision with which a pilot is able to perform those flight tasks required in support of an aircraft mission.(*146*) These qualities include not only the basic vehicle stability and control characteristics, but also the displays and controllers that comprise the pilot-vehicle interface.

Improbable: Conditions so unlikely that they are not anticipated to occur during the entire operational life of a single aircraft, but may occur several times during the operational life of the fleet. (*43*)

For military aircraft, conditions so unlikely that it can be assumed that occurrence will not happen during the lifetime of a single aircraft, but that it is possible within a fleet of a given type.(*43*) This corresponds to the civil term *Extremely Improbable*

Mission Analysis and Information Requirements (MAIR): A preliminary study and analysis undertaken prior to cockpit design to determine what information should be presented to the flight crew to accomplish the mission for which the aircraft and/or display is being designed. The MAIR will, to a large extent, form the basis for conducting the display evaluation testing.

Mission Task: A group of related flight tasks that, taken together, are representative of a portion of a mission.

Note: Examples of Mission Tasks are sling operations, pinnacle landings, or instrument approaches.

Mission Task Elements (MTE): A flight task that is representative of an element of a mission.

Note: This definition was changed from the ADS-33 definition.(*137*) The ADS-33 definition is that the MTE is an element of the mission. The definition was changed since the tasks are often highly stylized surrogate tasks that are similar to, but not normally included in a mission description. For example, the pirouette is an excellent example of a rotary-wing task that is similar to many found in low altitude, near earth operations, but does not exist in any mission statement.

Mode: The operational state of the display or flight control system: A selected group of display formats, input selections, and processing algorithms.

Operational Flight Envelope (OFE): The boundaries within which the aircraft is capable of operating in order to accomplish the operational missions.(*137*) These envelopes are defined in terms of combinations of airspeed, altitude, load factor, and any other parameters necessary.

Rating Level: Rating levels for handling qualities ratings (HQRs), for display readability ratings (DRRs), and for display flyability ratings (DFRs) are divided into four groups,
- Level 1: Satisfactory without improvement
- Level 2: Acceptable, deficiencies warrant, but do not require improvement
- Level 3: Unacceptable, deficiencies require improvement
- Level 4: Unacceptable, system is not usable.

Note: Level 4, Unacceptable/Unusable has been added for completeness.

See-Through Display: Any display through which the crew member may view the external real-world scene, normally, head-up or helmet-mounted displays.

Service Flight Envelope (SFE): The boundaries within which the aircraft is capable of operating.(*137*) These envelopes are defined in terms of same parameters as for the OFE, plus any other parameters necessary. The inner boundaries of the SFE are coincident with the outer boundaries of the OFE. The outer boundaries of the SFE are defined by one or more of the following: Uncommanded aircraft motion, or structural, engine, power-train, or rotor limits.

Situation Awareness (SA): The correct perception of the current operational environment, comprehension of the effect of the environment on vehicle status, and the projection of vehicle status to the future.(*87*)

Validation: The process of determining the degree to which a model is an accurate representation of the real world from the perspective of the intended uses of the model.(*46*)

Verification: The process of determining that a model's implementation accurately represents the developer's conceptual description and specifications.(*46*)

References

1   "Automated Cockpits Special Report. Part 1", *Aviation Week*, January 30, 1995, pp. 52-54

2   "Automated Cockpits Special Report. Part 2", *ibid.*, February 6, 1995, pp. 48-55

3   *Advanced Technology Aircraft Safety Survey Report*, Bureau of Air Safety Investigation, Canberra, June 1998, ISBN 0-642-27456-8

4   Sarter, N. B., and Woods, D. D., "How in the World Did We Ever Get into That Mode? Mode Errors and Awareness in Supervisory Control," *Human Factors*: 37, 1995, 5-19

5   Norman, D. A., *The Psychology of Everyday Things*, New York: Basic Books, 1988

6   "Indian A-320 Crash Probe Data Show Crew Improperly Configured Aircraft," *Aviation Week*, June 25, 1990, pp. 84-85

7   Billings, C. E., "Toward a Human-Centered Automation Philosophy," *Proceedings of the 5th International Symposium on Aviation Psychology, Columbus*, April 1989

8   Billings, C. E., *Human-Centered Aircraft Automation: A Concept and Guidelines*, NASA TM-103885, August 1991

9   Anderson, M. W., French, D. D., Newman, R. L., and Phillips, M. R., "Flight Testing a General Aviation Head-Up Display," *Journal of Aircraft*: 33, 1996, 235-238

10 *Airworthiness Standards: Normal, Utility and Acrobatic, and Commuter Category Airplanes*, Federal Aviation Regulations Part 23

11 *Airworthiness Standards: Transport Category Airplanes*, Federal Aviation Regulations Part 25

12 *United States Standard for Terminal Instrument Procedures (TERPS)*, FAA Handbook 8260.36

13 *Airborne Supplemental Navigation Equipment Using the Global Positioning System (GPS)*, FAA TSO-C129

14 Shafer, J. B., "Comment from the Editor," *Ergonomics in Design*, January 1996, p. 3

15 Ketchel, J. M., and Jenney, L. L., *Electronic and Optically Generated Aircraft Displays: A Study of Standardization Requirements*, JANAIR Report 680505, 1968

16 Singleton, W. T., "Display Design: Principles and Procedures," *Ergonomics*: 12, 1969, 519.531

17 Abbott, T. S., *Task-Oriented Display Design: Concept and Example*, SAE Paper 892230, 1989

18 Sexton, G. A., "Cockpit Crew Systems Design and Integration," *Human Factors in Aviation*, Wiener, E. L., and Nagel, D. C. (eds.), New York: Academic Press, 1988, pp. 495-526

19 Palmer, M. T., Rogers, W. H., Press, H. N., Latorella, K. A., and Abbott, T. S., *A Crew-Centered Flight Deck Design Philosophy for High Speed Civil Transport (HSCT) Aircraft*, NASA TM-109171, January 1995

20 Wilkins, R. R., *Designing the Conceptual Flight Deck for the Short Haul Civil Transport/Civil Tiltrotor*, SAE Paper 951997, 1995

21 Buchroeder, R. A., and Kocian, D. F., *Display System Analysis for the LHX Helicopter Application*, AAMRL TR-89-001, 1989

22 Newman, R. L., *Head-Up Displays: Designing the Way Ahead*, Aldershot, England: Ashgate Publishing, 1995

23 Haworth, L. A. and Newman, R. L., *Techniques for Evaluating Flight Displays*, NASA TM-103947, 1993

24 Newman, R. L., and Greeley, K. W., *Helmet-Mounted Display Design Guide*, AFDD TR-98-A-006, 1997

25 Rogers, S. P., and Myers, L. D., "Development of an Intelligent System to Air in Avionics Display Design", presented at *AIAA/IEEE Digital Avionics System Conference, Fort Worth*, October 1993

26 Storey, B. A., Rountree, M. E., Kulwicki, P. V., and Cohen, J. B., "Development of a Process for Cockpit Design," *Proceedings National Aerospace Electronics Conference (NAECON '94), Dayton*, 1994, pp. 688-695

27 J. Rasmussen, "Ecological Interface Design for Reliable Human Machine Systems," *International Journal of Aviation Psychology*: 9, 1999, 203-223

28 *Human Interface Design Methodology for Integrated Display Symbology*, SAE ARP-4155, 1995

29   McDaniel, J. W., "Obsolete Accounting Model Hinders Crew System Integration," *CSERIAC Gateway*: 6 [3], 1995, pp. 1-4

30   *Work Breakdown Structure for Defense Materiel Items (WBS)*, MIL-STD-881B, March 1993

31   Greeley, K. W. and Newman, R. L., *Development of a Cockpit Design Philosophy*, Crew Systems TR-95-15B, 1996

32   Greeley, K. W. and Newman, R. L., "Developing Integrated Cockpit Displays," *Free Flight, Proceedings 10th European Aerospace Conference, Amsterdam*, October 1997, paper 30

33   Salas, E., Prince, C, Bowers, C. A., Stout, R. J, Oser, R. L., and Cannon-Bowers, J. A., "A Methodology for Enhancing Crew Resource Management Training," *Human Factors*: 41, 1999, 161-172

34   Rolfe, J. M, "Human Factors and Flight Deck Design," *Aircraft Engineering*, November 1976, pp. 6-14, 22

35   Corker, K. M., and Smith, B. R., "An Architecture and Model for Cognitive Engineering Simulation Analysis: Application to Advanced Aviation Automation," Presented at the *AIAA Computing in Aerospace Conference, San Diego*, October 1993

36   Tamasi, G., and Pease, R. A., *MIDAS Evaluation of Proposed Civil Tilt Rotor Flight Deck Concepts*, SAE Paper 951996, 1995

37   Freund, L. E. and Sadosky, T. L., "Linear Programming Applied to Optimization of Instrument Panel and Workspace Layout," *Human Factors*: 9, 1967, 293-300

38   Bartlett, M. W., "Design of Control and Display Panels Using Computer Algorithms," *Human Factors*: 15, 1973, 1-7

39   Wickens, C. D., *The Proximity Compatibility Principle: Its Psychological Foundation and Its Relevance to Display Design*, Aviation Research Laboratory, University of Illinois Report ARL-92-5/NASA-92-3, September 1992

40   Andre, A. D., "Quantitative Layout Analysis for Cockpit Display Systems," *Society for Information Display International Symposium Digest of Technical Papers*: 23, 1992, pp. 647-650

41   Dinadis, N. and Vicente, K. J., "Designing Functional Visualizations for Aircraft Systems Status Displays," *International Journal of Aviation Psychology*: 9, 1999, 241-269

42   *System Safety Program Requirements*, MIL-STD-882C, January 1993

43   *System Design and Analysis*, FAA AC-25.1309-1A, June 1988

44 *Fault/Failure Analysis Procedure*, SAE ARP-0926A, November 1979

45 *Fault/Failure Analysis for Digital Systems and Equipment*, SAE ARP 1834, August 1986

46 Army Regulation 5-11

47 *Aircraft Display Symbology*, MIL-STD-1787B, April 1996

48 Fulmer, K. R., *Single Medium Flight Instrument Display Endorsement Process*, AFFSA White Paper, August 1998

49 *Human Factors Engineering Design Criteria for Helicopter Electro-Optical Display Symbology*, MIL-STD-1295A, June 1984; Canceled by Notice, May 1996

50 *Display, Head-Up, General Specification for*, MIL-D-81641(AS), June 1972

51 *Human Engineering Design Criteria for Military Systems, Equipment, and Facilities*, MIL-STD-1472D, March 1989

52 *Human Engineering Program Process and Procedures*, MIL-HDBK-46855A, May 1999

53 *Airworthiness Standards: Normal Category Rotorcraft*, Federal Aviation Regulations Part 27

54 *Airworthiness Standards: Transport Category Rotorcraft*, Federal Aviation Regulations Part 29

55 *Equipment, Systems, and Installations in Part 23 Airplanes*, FAA AC-23.1309-1C, March 1999

56 *Certification of Transport Category Rotorcraft*, FAA AC-29-2C, September 1999

57 *Certification of Normal Category Rotorcraft*, FAA AC-27-1A, July 1997

58 *Installation of Electronic Display Instrument Systems in Part 23 Airplanes*, FAA AC-23.1311-A, March 1999

59 *Transport Category Airplane Electronic Display Systems*, FAA AC-25-11, July 1987

60 "Airworthiness Guidance for Rotorcraft Instrument Flight," Appendix B to *Certification of Transport Category Rotorcraft*, FAA AC-29-2C, September 1999, (56)

61 *Flight Test Guide for Certification of Part 23 Airplanes*, FAA AC-23-8A, February 1989; Change 1, August 1993

62 *Flight Test Guide for Certification of Transport Category Airplanes*, FAA AC-25-7A, March 1998; Change 1 June 1999

63 *Automatic Pilot Systems Approval*, FAA AC 25.1329-1A, July 1968

64 *Minimum Flightcrew*, FAA AC-25.1523-1, February 1993

65 *Pilot Compartment View for Transport Category Airplanes*, FAA AC-25.773-1, January 1993

66 *Pilot Compartment View*, FAA AC-29.773-1, January 1966; This advisory circular may have been canceled.

67 *Category II Operations - General Aviation Airplanes*, FAA AC-91-16, August 1967; also *Category II Operations: Manual, Instruments, Equipment, and Maintenance*, FAR 91, Appendix A

68 *Criteria for Approving Category I and Category II Landing Minima for FAR 121 Operators*, FAA AC-120-29, September 1970; through Change 3, December 1974

69 *Criteria for Approval of Category III Landing Weather Minima for Takeoff, Landing, and Rollout*, FAA AC-120-28D, July 1999

70 *Automatic Landing Systems (ALS)*, FAA AC 20-57A, January 1971; Canceled by FAA AC-120-28D (*69*)

71 *Category 3 Operations with a Head-Up Display*, Joint Aviation Authorities JAR HUDS 901, April 1994

72 *Category 2 Operations with a Head-Up Display*, Joint Aviation Authorities JAR HUDS 902, April 1995

73 *Head-Up Displays*, Joint Aviation Authorities JAR HUDS 903, April 1995

74 *Minimum Performance Standards for Airborne Multipurpose Electronic Displays*, SAE AS-8034, December 1982

75 *Minimum Performance Standard for Airborne Head Up Display (HUD)*, SAE AS-8055

76 *Optical Measurement Procedures for Airborne Head Up Display (HUD)*, SAE ARP-5287

77 *Environmental Conditions and Test Procedures for Airborne Equipment*, RTCA DO-160D

78 *Defense System Software Development*, DoD-STD-2167A, 1988

79 *Software Considerations in Airborne Systems and Equipment Certification*, RTCA DO-178B, 1992

80 *Environmental Test Methods and Engineering Guidelines*, MIL-STD-810E, 1989

81 *Electronic Equipment, Airborne, General Specification for*, MIL-E-5400T, 1979

82 *Environmental and Test Procedures for Airborne Equipment*, RTCA DO-160D, 1997

83 *Requirements for the Control of Electromagnetic Emissions and Susceptibility*, MIL-STD-461D, 1993

84 *Electromagnetic Interference Characteristics, Measurements of*, MIL-STD-462

85 *Protection of Aircraft Electrical/Electronic Systems Against the Indirect Effects of Lightning*, FAA AC-20-136, 1990

86 St. Augustine, *Confessions*, ca. 397; translated by V. J. Bourke, Washington: Catholic University, 1953, p. 343

87 Jones, D. G. and Endsley, M. R., "Sources of Situation Awareness Errors in Aviation," *Aviation, Space, and Environmental Medicine*: 67, 1996, 507-512

88 Gawron, V. J., Weingarten, N. C., Hughes, T., and Adams, S., *Verifying Situational Awareness Associated with Flight*, AIAA Paper 99-1093, 1999

89 Besco, R. O., "What All Good Pilots Do: Does Anyone Know Or Care?", *Business and Commercial Aviation*, November 1996, pp. 108-114

90 *Trans World Airlines, Inc., Boeing 727-231, N54328, Berryville, Virginia, December 1, 1974*, NTSB AAR-75-16, November 1975

91 Coyle, S. and Krolak, W., *Common Sense Navigation for the 21st Century*, Transport Canada White Paper, 1999

92 *Descent Below Visual Glidepath and Collision with Terrain: Delta Air Lines Flight 554, MD-88, N914DL, LaGuardia Airport, New York, October 19, 1996*, NTSB AAR-97/03, August 1997

93 *Controlled Flight Into Terrain: American Airlines Flight 965, Boeing 757-223, Near Cali, Columbia, December 20, 1995*, Aeronautica Civil of the Republic of Columbia, Bogota, 1997

94 *NWA B-727-251, N274US, Near Thiells, New York, December 1, 1974*, NTSB AAR-75-13, August 1975; Ironically, this accident occurred on the same day as another situation awareness accident.(*90*)

95 Harvey, D. S., "V-22 Crash Prompts Questions About Cockpit Data And Displays," *Avionics*, August 1993, pp. 40-44

96 Ocker, W. C. and Crane, C. J., *Blind Flight in Theory and Practice*, San Antonio: Naylor Publishing, 1932

97 Benson, A. J., "Spatial Disorientation in Flight," in *Aviation Physiology*, Gillies, J. A. (ed.), London: Pergamon Press, 1965, pp. 1086-1129

98 *F-15 Spatial Disorientation*, Videotape produced by First Tactical Fighter Wing, Langley AFB, ca. 1986

99 *China Airlines Boeing 747SP, N-4522V, 300 Nautical Miles Northwest Of San Francisco, California, February 19, 1985*, NTSB AAR-86-03, March 1986; PB86-910405

100 Gawron, V. J., *SA Verification Guidance*, presentation to Tri-Service Flight Symbology Working Group, Brooks AFB, February 1999

101 Adams, S., "Practical Considerations for Measuring Situational Awareness," *Proceedings Third Annual Symposium on Situational Awareness in the Tactical Environment, Piney Point, Maryland*, June 1998, pp. 139-146

102 Mosier, K. L. and Chidester, T. R., "Situation Assessment and Situation in a Team Setting," in *Situational Awareness in Dynamic Systems*, Taylor, R. M. (ed.), RAF Institute of Aviation Medicine IAM Report 708, 1991

103 Endsley, M. R., "Situation Awareness Global Assessment Technique (SAGAT)," *Proceedings of the National Aerospace and Electronics Conference*, 1998, pp. 789-795

104 Endsley, M. R., "Design and Evaluation for Situation Awareness Enhancement," *Proceedings of the 32nd Annual Meeting, Human Factors Society*, 1998, pp. 97-101

105 Endsley, M. R., "Measurement of Situation Awareness in Dynamic Systems," *Human Factors*: 37, 1995, 65-84

106 Snow, M. P. and Reising, J. M., "Effect of Pathway-In-the-Sky and Synthetic Terrain Imagery on Situation Awareness in a Simulated Low-Level Ingress Scenario," *Proceedings Fourth Annual Symposium on Situational Awareness in the Tactical Environment, Piney Point, Maryland*, June 1999, pp. 198-207

107 Taylor, R. M., *Situational Awareness: Aircrew Constructs for Subject Estimation*, RAF Institute of Aviation Medicine IAM Report 670, 1990

108 Vidulich, M. A., and Hughes, E. R., "Testing a subjective metric of situation awareness," *Proceedings of the 35th Annual Meeting, Human Factors Society*, 1991, pp. 1307-1311

109 Carretta, T. R., Perry, D. C., and Ree, M. J., "Prediction of Situational Awareness in F-15 Pilots," *International Journal of Aviation Psychology*: 6, 1996, 21-41

110 Newman, R. L., *Evaluation Of Head-Up Displays To Enhance Unusual Attitude Recovery*, AFWAL TR-87-3055, Vol. II, June 1987

111 Bailey, R. E. and Knotts, L. H., *Flight And Ground Simulation Evaluation Of The Proposed USAF Head-Up Display Standard*, AIAA paper 93-3605, August 1993

112 Penwill, J. C. and Hall, J. R., *A Comparative Evaluation Of Two HUD Formats By All Four Nations To Determine The Preferred Pitch Ladder Design For EFA*, RAE FM-WP(90)021, March 1990

113 Newman, R. L., *Helmet-Mounted Display Flight Symbology Requirements for Military Rotorcraft: Background Document*, TR-98-02A, October 1999

114 Gallimore, J. J., Brannon, N. G., Patterson, F. R., and Nalepka, J. P., "Effects of FOV and Aircraft Bank on Pilot Head Movement and Reversal Errors During Simulated Flight," *Aviation, Space, and Environmental Medicine*: 70, 1999, 1152-1160

115 Gawron, V. J., *SA Verification Guidance*, presentation to Tri-Service Flight Symbology Working Group, Brooks AFB, February 1999

116 Kelley, C. R. *Manual and Automatic Control*, New York: Wiley, 1968

117 Birmingham, H. P. and Taylor, F. V., "A Design Philosophy for Man-Machine Control Systems," *Proceedings IRE*: 42, 1954, 1748-1758; reprinted in Sinaiko, H. W. (ed.) *Selected Papers on Human Factors and Use of Control Systems*, New York: Dover, 1961, pp. 67-87

118 Gawron, V. J., *SA Verification Guidance*, presentation to Tri-Service Flight Symbology Working Group, Brooks AFB, February 1999

119 Obermeyer, R. W. and Vreuls, D., *Data Collection And Measurement Techniques For Manned Flight System Studies*, Bunker-Ramo Paper, July 1967

120 Wierwille, W. W., and Connor, S. A., "Evaluation of 20 Workload Measures Using a Psychomotor Task in a Moving-Base Simulator," *Human Factors*: 25, 1983, 1-16

121 Hubbard, D. C., Rockway, M. R., and Waag, W. L., "Aircrew Performance Assessment," *Aviation Psychology*, Jensen, R. S. (ed.), Aldershot: Gower Technical, 1989, pp. 342-377

122 Dunn, R. S. and McBride, D., "Frequency Weighted Task Complexity Index, and Improved Metric for Man-Machine Integration," Presented to *DoD Human Factors Engineering Technical Advisory Group Meeting 42*, May 1999

123 Johnson, S. L. and Roscoe, S. N, "What Moves, The Airplane or the World?", *Human Factors*: 14, 1972, 107-129

124 Roscoe, S. N. and Williges, R. C, "Motion Relationships in Aircraft Attitude and Guidance Displays," *Human Factors:* 17, 1975, 374-387

125 Schifflett, S., *Evaluation of a Pilot Workload Assessment Device to Test Alternate Display Formats and Control Handling Qualities*, NATC SY-33R-80, July 1980

126 Newman, R. L. and Bailey, R. E., *Head-Up Display ILS Accuracy Flight Test*, AFWAL 87-3015-V, June 1987

127 Kantrowitz, B. H. and Casper, P. A., "Human Workload in Aviation," *Human Factors in Aviation*, Weiner, E. L. and Nagel, D. C. (eds.), New York: Academic Press, 1988, pp. 157-187

128 Charlton, S. G., "Mental Workload Test and Evaluation," *Handbook of Human Factors Testing and Evaluation*, O'Brien, T. G. and Charlton, S. G. (eds.), Mahwah, NJ: Erlbaum, 1996, pp. 181-199

129 Wierville, W. W. and Casali, J. G., "A Validated Rating Scale for Global Mental Workload Measurement Application," *Proceedings 27th Annual Meeting Human Factors Society*, 1983, pp. 120-133

130 Reid G. B. and Nygren, T. E., "The Subjective Workload Assessment Technique: A Scaling Procedure for Measuring Mental Workload," *Human Mental Workload*, Hancock, P. A. and Meshkati, N. (eds.), Amsterdam: North Holland, 1988, pp. 185-218

131 Vidulich, M. A., Ward, G. F., and Schueren, J., "Using the Subjective Workload Dominance (SWORD) Technique for Projective Workload Assessment," *Human Factors*: 33, 1991, 677-691

132 Hart, S. G. and Staveland, L. E., "Development of the NASA Task Load Index (TLX): Results of Empirical and Theoretical Research," *Human Mental Workload*, Hancock, P. A. and Meshkati, N. (eds.), Amsterdam: North Holland, 1988, pp. 139-183

133 *Task Load Index*, NASA Ames Research Center, Human Performance Research Group, ca. 1987

134 Spyker, D. A., Stackhouse, S. P., Khalafalla, A. S., McLane, R. C., *Development of Techniques for Measuring Pilot Workload*, NASA CR-1888, November 1971

135 Garman, P. J., and Trang, J. A., "In Your Face! The Pilot's/Tester's Perspective on Helmet-Mounted Display (HMD) Symbology," *Proceedings Helmet- and Head-Mounted Displays and Symbology Requirements Symposium, Orlando*, April 1994, *Proceedings SPIE*: 2218 pp. 274-280

136 Haworth, L. A., Gillow, C., and Newman, R. L., "A Standardized Diagnostic Method for Separating Helmet-Mounted Display Symbology Dynamics," *SETP Cockpit*, Second Quarter 1995, pp. 6-17

137 *Handling Qualities Requirements for Military Rotorcraft*, ADS-33D-PRF, 1996

138 *Flying Qualities of Piloted Airplanes*, MIL-F-8785C, 1979

139 *Instrument Flying*, AFI 21-216; formerly AFM-51-37

140 *All-Weather Flight Manual*, NAVAER 80-80T-60, 1957

141 *Instrument Flight Handbook*, FAA AC-61-27C, November 1979

142 *Aircrew Training Manual, Attack Helicopter, AH-1*, TC 1-213, 1992

143 Wright, B. A., and Hoobler, M. A., *Standardized Head-Up Display Symbology Evaluation*, AFFTC TR-91-04, November 1991

144 *Quality Assurance of Software Used in Aircraft or Related Products*, FAA AC-21-33, February 1993

145 Newman, R. L. and Anderson, M. W., "HUD Flight Testing: Lessons Learned," presented at *Southeast Section SETP Symposium, Stone Mountain, Georgia*, May 1994

146 Cooper, G. E., and Harper, R. P., *The Use of Pilot Rating in the Evaluation of Aircraft Handling Qualities*, NASA TN-D-5153, 1969

147 Hoh, R. H., Mitchell, D. G., Aponso, B. L., Key, D. L., and Blanken, C. L., *Background Information and User's Guide for Handling Qualities Requirements for Military Rotorcraft*, USAAVSCOM TR-89-A-008, December 1989

148 Chiappetti, C. F., *Evaluation of the Haworth-Newman Avionics Display Readability Scale*, Thesis: Naval Postgraduate School, September 1994

149 Hoh, R. H. *Investigation of Outside Visual Cues Required for Hover*, AIAA Paper 85-1808, 1985

150 Burgess, M. A., Chang, T., Dunford, D. E., Hoh, R. E., Horne, W. F., and Tucker, R. F., *Synthetic Vision Technology Demonstration, Executive Summary*, DOT/FAA/RD-93/40, Volume 1, December 1993

151 Burgess, M. A., Chang, T., Dunford, D. E., Hoh, R. E., Horne, W. F., Tucker, R. F., and Zak, J. A., *Synthetic Vision Technology Demonstration, Flight Tests*, DOT/FAA/RD-93/40, Volume 3, December 1993

152 *Aircraft Internal Time Division Command/Response Multiplex Data Bus*, MIL-STD-1553, September 1978

153 *Digital Information Transfer System (DITS)*, Aeronautical Radio, Inc. Specification ARINC-429

154 *Ground Minimum Performance Standards, Airborne Ground Proximity Warning Equipment*, RTCA DO-161A, May 1976

155 *Minimum Performance Standards for Traffic Alert and Collision Avoidance System (TCAS)*, RTCA DO-195

156 Spitzer, C. R., Digital Avionics Systems, 1987, Englewood Cliffs, NJ: Prentice-Hall

157 Newman, R. L., Schwartz, R. J., Greeley, K. W., and Ellis, D. R., *An Affordable General Aviation Head-Up Display*, Crew Systems Report, TR-98-34 to NASA for Contract NAS1-97148, August 1998

158 Herrington, R. M., Shoemacher, P. E., Barlett, E. P., Dunlap, E. W., *Flight Test Engineering Handbook*, Air Force TR-6273, May 1951; revised January 1966

159 O'Brien, T. G., "Anthropometry, Workspace, Environmental Test and Evaluation," *Handbook of Human Factors Testing and Evaluation*, O'Brien, T. G. and Charlton, S. G. (eds.), Mahwah, NJ: Erlbaum, 1996, pp. 223-264

160 Van Cott, H. P. and Kinkade, R., *Human Engineering Guide to Equipment Design*, Washington: American Institute for Research, 1972

161 *Approval Basis for Automatic Stabilization Equipment (ASE) Installations in Rotorcraft*, FAA AC-29-1, December 1963

162 R. L. Newman, *Helmet-Mounted Display Symbology and Stabilization Concepts*, NASA CR-196697, June 1995

163 Markman, S. R., *USAF In-Flight Simulation: A Cost Effective Operating Approach*, AIAA Paper 93-3604, August 1993

164 The Safety Assessment of Systems, British Civil Airworthiness Requirements Paper 670, May 1979

Index

For Product Safety Concerns and Information please contact our
EU representative GPSR@taylorandfrancis.com Taylor & Francis
Verlag GmbH, Kaufingerstraße 24, 80331 München, Germany